Current Topics in Microbiology 147 and Immunology

Oncogenes

Selected Reviews

Edited by P. K. Vogt

With 8 Figures

Springer-Verlag
Berlin Heidelberg NewYork
London Paris Tokyo Hong Kong

PETER K. VOGT, Ph. D.

Department of Microbiology
University of Southern California
School of Medicine
2011 Zonal Avenue HMR-401
Los Angeles, CA 90033-1054
USA

ISBN-13: 978-3-642-74699-4 e-ISBN-13: 978-3-642-74697-0
DOI: 10.1007/978-3-642-74697-0

Library of Congress Catalog Card Number 15-12910
Softcover reprint of the hardcover 1st edition 1989

Offsetprinting: Saladruck, Berlin; Bookbinding: Helm, Berlin
2123/3020-543210 — Printed on acid-free paper

Table of Contents

Indexed in Current Contents

List of Contributors

You will find their addresses at the beginning of the respective contributions

Boettiger, D.
Broek, D.
Hoffmann, F. M.
Holland, G. D.
Parsons, J. D.
Risser, R.
Weber, M. J.

Roles of *Drosophila* Proto-oncogene and Growth Factor Homologs During Development of the Fly

F. MICHAEL HOFFMANN

1 Introduction

The regulation of cell proliferation and differentiation is an essential aspect of organismal development. An important goal of investigations over the next few years will be to define the regulatory pathways of cellular proliferation and differentiation by discovering how the molecules we know about, and others that remain to be discovered, interact with one another to instruct a cell to divide or to differentiate. Growth factors, growth receptors, and the proto-oncogenes represent some of the components on these regulatory circuits; the challenge is to complete the "wiring-diagrams" by finding the molecular connections between the components. Molecular genetic and biochemical approaches are being applied to this problem in vertebrate cell culture and animal systems. The opportunity to use classical genetic strategies to address this problem led several groups to examine simpler organisms for the presence of proteins homologous to growth factors, growth factor receptors, or proto-oncogenes. It is now clear that many of these proteins have been well conserved during evolution and are found in three organisms eminently suited to genetic manipulation—yeast, *Caenorhabditis elegans*, and *Drosophila melanogaster*.

McArdle Laboratory for Cancer Research, University of Wisconsin, Madison, WI 53706, USA

Current Topics in Microbiology and Immunology, Vol. 147

The goal of this review is to provide some perspective on the range of processes involving these genes in the development of a metazoan. The examples described here represent the first *Drosophila* proto-oncogene and growth factor homologs for which the availability of mutant alleles has provided some information about their roles in development. The number of *Drosophila* genes identified as homologous to proto-oncogenes, growth factors, or growth factor receptors will undoubtedly increase greatly over the next few years. There are several reasons for interest in the *Drosophila* homologs. The first is the opportunity to isolate mutations in a specific gene and study the phenotypic ramifications on normal cellular functions and developmental processes in the context of the whole organism. One can ask: What is the effect of eliminating a specific gene product on the developmental program of the organism? *Drosophila* genetics can also be used to identify genes whose products interact with a specific proto-oncogene, growth factor, or growth factor receptor in its regulatory pathway. Mutations in a proto-oncogene, for example, can be used to contrive a genetic background upon which it is possible to identify second-site mutations that either enhance or suppress the mutant phenotypes caused by the lesions in the proto-oncogene. This approach has the potential for identifying regulators and substrates of the proto-oncogene and its gene product. A third reason for the interest in the *Drosophila* homologs stems from the fact that regulation of cell proliferation and differentiation are important aspects of *Drosophila* development. These homologous genes provide identifiable molecules that may participate in these aspects of *Drosophila* development, i.e., they provide inroads to potentially interesting areas of *Drosophila* developmental biology.

The *Drosophila* genes have been identified either by actively screening for the *Drosophila* genes with vertebrate probes under low stringency conditions or by serendipitously finding similarity between the amino acid sequence of an identified *Drosophila* gene and the amino acid sequence of a vertebrate gene. While many sequences have been identified (Shilo 1987), mutant alleles are currently available only for the nine genes discussed here. Each of these genes will be described in the order of their earliest activity during *Drosophila* development, beginning with oogenesis. As we proceed through the course of development, later roles for the genes will be noted based on our current understanding of mutant phenotypes and the localization of gene products.

2 Oogenesis and Maternal Effects on Embryogenesis

The *Drosophila* embryo accomplishes the progression from a fertilized egg to a highly differentiated, mobile, larval form in a period of 24 h. This process requires the controlled proliferation and migration of thousands of cells and the terminal differentiation of these cells into specialized tissues. Many of the gene products required to initiate and carry out this rapid process are placed in the egg during oogenesis. Several genes have been identified as having maternal effects on early *Drosophila* embryogenesis, affecting such basic processes as the determination of position along the anterior/posterior axis (Schüpbach and Wieschaus 1986;

NÜSSLEIN-VOLHARD et al. 1987) and the determination of position along the dorsal/ventral axis (ANDERSON and NÜSSLEIN-VOLHARD 1984) in the embryo. The larval cuticle formed during embryogenesis is extremely useful for the analysis of pattern formation during *Drosophila* embryogenesis in that it has characteristic structures demarcating each of the segments of the animal along the anterior/posterior axis as well as unique structures along the dorsal/ventral axis of the embryo.

The *Drosophila* oocyte develops through the cooperation of germline and somatic cells (MAHOWALD and KAMBYSELLIS 1980). A germline stem cell gives rise after four mitotic divisions to a syncytium of 15 nurse cells and one oocyte. The cells are connected by cytoplasmic bridges through which RNAs and proteins synthesized in the nurse cells are transported to the oocyte. Specialized somatic cells, called follicle cells, envelop the syncytium and are responsible for the transport of yolk proteins from the hemolymph to the oocyte and for the secretion of the proteins that make up the vitelline membrane and the chorion of the eggshell. Genes have been identified that function either in the follicle cells or the nurse cells to produce the asymmetrical structures of the *Drosophila* eggshell; the function of some of these genes during oogenesis affects not only the pattern of the eggshell but also the patterns in the embryo that develop from the egg (reviewed by IRISH 1987).

In the following sections, several *Drosophila* homologs with critical roles during oogenesis will be described. Normal function of the *Drosophila* epidermal growth factor (EGF) receptor homolog is required during oogenesis in the somatically derived follicle cells for proper pattern formation of the eggshell and of the embryo (Sect. 2.1). Expression of the *Drosophila* proto-oncogene homologs of c-*rel*, c-*abl*, and c-*raf* is required during oogenesis in the germline-derived nurse cells for proper development of the embryo (Sect. 2.2–2.4).

2.1 *faint little ball/torpedo*: An EGF Receptor Homolog

The *Drosophila* EGF receptor homolog DNA was identified by cross-hybridization to the chicken oncogene, v-*erb*B (LIVNEH et al. 1985; WADSWORTH et al. 1985). v-*erb*B is one of two oncogenes in avian erythroblastosis virus and is derived from the proto-oncogene encoding the EGF receptor (DOWNWARD et al. 1984). Although the human genome contains two similar genes, EGF receptor and the c-*neu* or *erb*B-2 proto-oncogene (SCHECHTER et al. 1984), the *Drosophila* genome contains only one homolog, as indicated by the failure of the *Drosophila* probe to recognize more than one Drosophila sequence under low stringency conditions. The *Drosophila* DER sequence is equally similar to the human EGF receptor and c-*neu* proto-oncogenes (SCHEJTER et al. 1986).

The DNA sequence of the *Drosophila* clones was determined and used to derive the amino acid sequence of a predicted protein product (SCHEJTER et al. 1986). The *Drosophila* protein has several regions of homology with the human EGF receptor. The putative intracellular portion of the protein includes a 309 amino acid domain with 55% amino acid similarity to the tyrosine kinase domain of the human EGF receptor. The *Drosophila* protein contains a putative transmembrane domain of 31 hydrophobic amino acids in a position that aligns with the transmembrane domain of the human protein. Sequences on the two alternate 5′ exons encode hydrophobic amino acid sequences that presumably act as transmembrane signal sequences. The reported existence of a third alternative 5′ exon (SCHEJTER et al. 1986) is an error due to a cloning artifact (B. Z. SHILO, personal communication). The *Drosophila* sequence contains three cysteine-rich regions in an extracellular domain of 769 amino acids with 37% amino acid similarity to the extracellular domain of the human EGF receptor, and 33% amino acid similarity to the extracellular domain of the human c-*neu* proto-oncogene (SCHEJTER et al. 1986). Thus far, a ligand for the *Drosophila* receptor has not been identified. Mammalian EGF and transforming growth factor (TGF)-α, two agonists for the vertebrate EGF receptor, do not enhance the tyrosine kinase activity of the *Drosophila* protein (B.-Z. SHILO, personal communication).

Two transcripts of 7.6 and 7.1 kilobases (kb) (LEV et al. 1985) or 6.6 and 6.3 kb (KAMMERMEYER and WADSWORTH 1987) are expressed at a variety of times during development and in several different tissues (Sect. 3.4, 4.3). During oogenesis, follicle cells express higher levels of the RNA than the adjacent nurse cells and oocytes (KAMMERMEYER and WADSWORTH 1987). The RNA is observed in the follicle cells at early stages in oogenesis but not in the follicle cells that surround mature oocytes; perhaps the nonproliferating follicle cells no longer require the receptor RNA (KAMMERMEYER and WADSWORTH 1987).

The *Drosophila* EGF receptor homolog is encoded by the previously described *Drosophila* gene *faint little ball* (*flb*) (NÜSSLEIN-VOLHARD et al. 1984; PRICE et al. 1989; SCHEJTER and SHILO, 1989). The zygotic mutant phenotypes of *flb* will be described in Sect. 3.4; however, a subset of *flb* mutant alleles, originally named *torpedo* (*top*) (SCHÜPBACH 1987), retainsufficient *flb* functions to permit the mutant animals to reach adulthood. *top* mutations are female sterile mutations that affect the structure of the eggshell secreted by the follicle cells and the polarity of the dorsal/ventral pattern in the embryonic offspring (SCHÜPBACH 1987). Genetic analysis and manipulation of the germline precursor cells, the pole cells, were used to demonstrate that *top* expression in the somatically derived follicle cells of the ovary is essential for normal eggshell development and for the subsequent embryonic development of the progeny. The specific role of the receptor protein in the follicle cells is under investigation; however, it seems likely that it may be required for optimal proliferation of the follicle cells during oogenesis or for mediating communication between the follicle cells and the oocyte that leads to specific pattern elements of the eggshell. The distinct patterns of expression of specific chorion genes (PARKS and SPRADLING 1987) suggest that the follicle cells may be receiving positional signals from the oocyte. There are positional asymmetries present in the oocyte, e.g., the location of the oocyte nucleus, which could provide positional information to the follicle cells. Communication between the oocyte and the follicle cells is also implied by the observation that the *FS(1)K10* gene, which is expressed by the oocyte nucleus, imparts a mutant phenotype on the pattern of the eggshell produced by the follicle cells (HAENLIN et al. 1987). *flb*/*top* may be involved in mediating this cell-to-cell communication between cells of the germline and the soma.

2.2 *abl*: A Cytoplasmic Tyrosine Kinase Homolog

The *Drosophila* Abelson proto-oncogene homolog (*abl*) was identified by cross-hybridization with a v-*abl* oncogene probe (Hoffman-Falk 1983; Hoffmann et al. 1983). The *abl* cytoplasmic tyrosine kinase was originally identified as the activated oncogene of the Abelson murine leukemia virus and has since been implicated in the oncogenesis of human chronic myelogenous leukemia (Witte 1986; reviewed by Risser and Holland, this volume). The *Drosophila* gene consists of 10 exons distributed across 26 kb of DNA with an open reading frame of 1521 codons (Henkemeyer et al. 1988). Extensive amino acid sequence similarity (80%) between the *Drosophila* and human *abl* genes begins at the amino terminal glycine residue and extends through the three domains characteristic of the cytoplasmic tyrosine kinase family: the SH3, the SH2 (Sadowski et al. 1986), and the tyrosine kinase catalytic domains (Henkemeyer et al. 1988). The SH2 and SH3 regions are also present in the phosphatidyl inositol-specific phospholipase C and the v-*crk* oncogene (Mayer et al. 1988; Stahl et al. 1988) and may serve as regulatory domains of the cytoplasmic tyrosine kinases and the phospholipase. Distortion of these regulatory mechanisms by the v-*crk* oncoprotein could be the basis of its oncogenic activity.

abl sequences also have been isolated from *Calliphora erythrocephala* by cross-hybridization with *Drosophila abl* probes (Durica et al. 1987) and from the nematode *Caenorhabditis elegans* by cross-hybridization to a v-*src* probe under low stringency conditions (Goddard et al. 1986). Only the portion of the *Calliphora* gene corresponding to exons 3 and 4 of the *Drosophila abl* gene (the kinase catalytic domain) has been sequenced; the *Calliphora* sequence is virtually identical with the *Drosophila* sequence, as is the organization of the gene as determined by the position of introns within the sequence. The amino acid sequence similarity between the nematode sequence and v-*abl* is 62% through the SH2 and tyrosine kinase domains (Goddard et al. 1986).

As in the case of *flb*, the *abl* RNA is expressed at several times and in several tissues during development; however, the levels of the 6.0- to 6.5-kb transcript from the *Drosophila abl* gene are highest in unfertilized eggs (Lev et al. 1984; Wadsworth et al. 1985; Telford et al. 1985; K. D. Rand and F. M. Hoffmann, unpublished observations). The maternally provided *abl* RNA is distributed uniformly throughout the preblastoderm embryo, and as the syncytial blastoderm forms, the transcript becomes restricted to the cortical material directly below the nuclei as they reach the plasma membrane (M. J. Henkemeyer and F. M. Hoffmann, unpublished observations). As soon as gastrulation begins, as indicated by the formation of the ventral furrow, *abl* RNA becomes undetectable. *abl* RNA and protein are detected later in development in differentiating nerve cells (see Sect. 3.4, 4.3).

Mutations in the *Drosophila abl* gene were generated by isolating lethal mutations in an 18-gene region uncovered by a chromosomal deletion. The *abl* alleles were identified among the induced lethals by their lack of genetic complementation with chromosomal lesions that affect the *abl* transcriptional unit and by rescue of their mutant phenotypes with *abl* DNA in transgenic *Drosophila* (Henkemeyer et al. 1987; reviewed by Hoffmann et al. 1989). Mutations in the *abl* gene are lethal to the zygote (Sect. 4.2); therefore, to determine whether the *abl* transcript in the oocyte has an essential function, it is necessary to produce clones of germline cells, mutant for *abl*, in females that are otherwise wild type for *abl* gene function. These clones of mutant cells are produced by radiation-induced mitotic recombination in the germline precursor cells (PERRIMON and GANS 1983). Examination of embryos derived from these *abl* null germline clones indicates that the maternally contributed *abl* transcript is required for normal embryogenesis (F. M. HOFFMANN, unpublished observations). The cuticles formed in embryos derived from oocytes that are mutant for the *abl* gene show severe loss of structures, indicating that the *abl* tyrosine kinase is probably required for proper cellularization and/or viability of cells in the early blastoderm.

2.3 *dorsal*: A c-*rel* Homolog

The *dorsal* gene (*dl*) was cloned by using chromosomal inversions with breakpoints in the *dl* gene and in a cloned region near the *dopa decarboxylase* gene (STEWARD et al. 1984). The DNA sequence of *dl* predicts a protein product of 677 amino acids with amino acid sequence similarity to the proto-oncogene c-*rel*. The region of homology extends over 295 amino acids and shows 47% identity and 80% similarity to the c-*rel* sequence (STEWARD 1987). The c-*rel* proto-oncogene was discovered as the normal cellular counterpart of the v-*rel* oncogene found in reticuloendotheliosis virus strain T, Rev-T (WILHELMSEN et al. 1984). The v-*rel* protein contains a nuclear localization signal and has been found in the nucleus (GILMORE and TEMIN 1988); however, there is no evidence that the protein is a DNA-binding protein. One hypothesis is that it may participate in the formation of protein complexes, perhaps even transcription initiation complexes, by virtue of protein/protein interactions with proteins that bind DNA directly (T. GILMORE, personal communication).

A 2.8-kb *dl* RNA is present in RNA from dissected ovaries and early embryos, but not in RNA from later embryonic or other stages of *Drosophila* development (STEWARD et al. 1985; STEWARD et al. 1988). The level of *dl* RNA remains constant through the first ten nuclear divisions in the embryo and disappears from the embryo at the time of cellular blastoderm formation. This pattern of expression is expected of a gene that is expressed during oogenesis and has its sole function in early embryogenesis. In situ hybridization to *dl* mRNA indicates that the RNA is found in the nurse cells of the ovary and uniformly distributed in the oocyte and early embryo. In contrast to the uniform distribution of *dl* RNA, the *dl* protein is detected in the nuclei only along the ventral aspect of the embryo (STEWARD et al. 1988). In cross-sections of the embryos, a gradient of *dl* protein can be observed, with the highest amount seen in the ventralmost nuclei. These are the cells of the blastoderm that will give rise to the mesoderm during gastrulation. This ventral pattern of *dl* protein is maintained, but the level of protein falls early in gastrulation, and the protein is undetectable by the time of germ band extension. The uniform distribution of the *dl* transcript and asymmetric distribution of *dl* protein suggest that there may be differential translation of the *dl* message dependent on the position of that message in the embryo or that the *dl* protein is unstable and rapidly degraded at dorsal positions in the animal.

dl function is required maternally to establish dorsal/ventral polarity in the early embryo (Nüsslein-Volhard 1979). Females that are homozygous for a *dl* mutation produce embryos that are defective in the ventralmost morphogenetic movements of gastrulation and do not differentiate ventral cuticular structures. The earliest observed defect is a failure of ventral furrow formation, the initial morphogenetic movement of gastrulation. The blastoderm cells that normally invaginate during ventral furrow formation give rise to mesodermal structures in the embryo, and in the absence of ventral furrow formation these tissues fail to form. The cuticle formed from an embryo derived from a *dl* mutant mother bears only dorsal cuticular structures and none of the structures characteristic of ventral positions on the animal. The absence of ventral cuticular structures and the failure of the ventral furrow to form are both consistent with the view that the information that specifies ventral position to the cells of the blastoderm is missing in *dl* mutant embryos. The *dl* gene apparently has no later functions during *Drosophila* development based on its genetic properties as a strict maternal-effect mutation and the absence of *dl* transcripts at any other stage in development.

The *dl* gene is one of 12 maternal-effect genes that have been identified as being required during oogenesis for proper dorsal/ventral pattern formation in the embryo (Anderson 1987). The molecular interactions between these genes are not known, but several other homologies to proteins of known function have been found among this group. For example, the *snake* and *easter* genes encode proteins with homologies to serine proteases (DeLotto and Spierer 1986; K. V. Anderson, personal communication). The localization of the *dl* protein to the nuclei along the ventral surface of the *Drosophila* embryo is consistent with a role in specifying events related to ventral furrow formation and in triggering determinative events that later provide cells with the ability to secrete cuticular structures appropriate for the ventral surface of the animal. The gradient of *dl* protein could be involved in activating ventrally active zygotic genes or in repressing dorsally active zygotic genes in the embryo. For example, the *twist* gene is expressed zygotically in the mesoderm (Thisse et al. 1987), and the *dl* protein might participate in the activation of this gene. On the other hand, *zen* and *dpp* are initially expressed along the dorsal aspect of the animal (Rushlow et al. 1987; St. Johnston and Gelbart 1987), and *dl* protein could be involved in repressing the expression of these genes in more ventral regions. Consistent with this latter model is the observation that both *zen* and *dpp* are expressed uniformly in *dl* mutant animals (Rushlow et al. 1987, D. N. Gunn and F. M. Hoffmann, unpublished observations).

2.4 *pole hole*: A c-*raf* Homolog

Drosophila raf proto-oncogene homologs were identified by cross-hybridization to a murine v-*raf* oncogene (Mark et al. 1987) and, independently, by cross-hybridization to the human *raf* oncogene (Nishida et al. 1988). The v-*raf* oncogene encodes a serine/threonine-specific protein kinase (Moelling et al. 1984) and is present in the murine sarcoma virus 3611 and the avian sarcoma virus MH2 (v-*mil*) (Rapp et al. 1983; Jansen et al. 1983). The *raf* gene in human cells may be activated in stomach cancer and in some glioblastomas (Shimizu et al. 1985;

FUKUI et al. 1985). Two *Drosophila* sequences, *Draf-1* and *Draf-2*, were identified by low stringency cross-hybridization with a probe spanning the kinase domain of v-*raf*. Detailed characterization of the *Draf*-2 locus has not been carried out, although it may represent the *Drosophila* equivalent of the human *raf*-related gene, *PKS* (MARK et al. 1986).

Complete sequence analysis of a 4.3-kb genomic fragment and complementary (c)DNA fragments of *Draf-1* revealed an open reading frame encoding a protein of 666 amino acids (NISHIDA et al. 1988). The sequence exhibits 45% amino acid identity with the human sequence over the entire protein, and 65% identity in the carboxy terminal 305 amino acids, corresponding to the serine/threonine-specific protein kinase domain. Four other blocks of sequence with high degrees of similarity may represent domains that have important functions in the activity of the protein. The *Draf-1* RNA of 2.9 kb is expressed in the ovary, as well as in other tissues (Sect. 4.3), consistent with a maternal contribution to early embryogenesis (NISHIDA et al. 1988).

A 4.3-kb genomic fragment was cloned into a P-element transformation vector for germline transformation. Two of the 11 mutations in the chromosomal region 2F1 through 3A4 were rescued by the presence of the transposon and defined as mutant alleles (NISHIDA et al. 1988). The mutations in the *Draf-1* sequence are alleles of a previously known gene called l(1)*pole hole* (N. PERRIMON and Y. NISHIDA, personal communication). Examination of the function of the *pole hole* maternal transcript was accomplished by inducing clones of *pole hole* mutant cells in the germline of the heterozygous females by mitotic recombination (PERRIMON et al. 1985; NISHIDA et al. 1988). The results of the germline clone analysis of *pole hole* indicate that the gene is not required for oogenesis, but that the maternal transcript can affect embryogenesis. The embryos derived from the *pole hole* mutant oocytes show two distinguishable mutant phenotypes depending on whether the zygote inherits a wild-type or mutant *pole hole* allele from the father. Embryos from *pole hole* mutant oocytes that also inherit a mutant *pole hole* allele from the father show severe defects at the cellular blastoderm stage consisting of disorganized layers of uncellularized nuclei. These embryos do not differentiate recognizable cuticular structures. In contrast, embryos from *pole hole* mutant oocytes that inherit a wild-type *pole hole* allele from the father have a fairly normal cellular blastoderm with defects in cellularization only at the posterior pole, just beneath the progenitor cells of the germline, the pole cells. These embryos develop with defects specifically at the posterior pole involving loss of posterior cuticular structures, including the spiracles, filzkorper, and anal tufts. The alleviation of the mutant phenotype by the paternally inherited wild-type *pole hole* gene suggests that both maternal and early zygotic expression of *pole hole* are important for cellular blastoderm formation. Other functions of *pole hole* later in development are described in Sect. 4.3.

3 Embryonic Zygotic Expression: Effects on Pattern Formation and Organogenesis

While the critical events of the first 2 h of embryogenesis (establishment of positional asymmetry, nuclear multiplication, migration, and cellularization) are largely carried out by gene products supplied to the oocyte maternally, de novo gene expression in the zygote plays a critical role in the earliest manifestations of pattern formation at the cellular blastoderm stage. Zygotic gene expression does not begin until the later stages of the syncytial blastoderm, shortly before cellularization of the blastoderm. Some of the genes expressed at this time have critical roles in setting up the basic body plan of the animal. They participate in the sequential subdivision of the blastoderm to a segmented embryo, in the initial morphogenetic movements of gastrulation that give rise to the embryonic germ layers, and in the determination of specific cell types (reviewed by SCOTT and CARROLL 1987; INGHAM 1988). The process of gastrulation begins with the invagination of cells along the ventral midline to form the ventral furrow (CAMPOS-ORTEGA and HARTENSTEIN 1985). A coordinate migration of cells begins at the posterior end of the animal where the cells form a flat surface called the dorsal plate, to which the precursors to the germline, the pole cells, adhere. The cells of the dorsal plate then move dorsally and anteriorly to produce the amnioproctodeal invagination. The majority of the cells of the blastoderm migrate ventrally as ventral furrow invagination proceeds to produce the germ band consisting of cells of the dorsal ectoderm, the ventral neurectoderm, and the mesoderm. The germ band extends in the major morphogenetic movement of embryogenesis around the posterior end of the animal and across the dorsal surface to form a U-shaped band of cells. The germ band then retracts, or shortens, back onto the ventral surface of the embryo, and the cells of the hypoderm migrate dorsally in the process of dorsal closure. Organogenesis takes place during the second half of embryogenesis, culminating in the hatching of the larva.

Even among the small number of homologs being considered here, examples exist of genes involved in each of the major developmental processes in the early zygote: determination, proliferation, and organogenesis. The *decapentaplegic* and *wingless* genes participate in pattern formation during embryogenesis in the dorsal/ventral and anterior/posterior directions respectively (Sect. 3.1, 3.2). The *flb* gene is required for embryonic cell proliferation or viability (Sect. 3.4). *Notch*, *Delta*, and *abl* all participate in aspects of nervous system development in the embryo (Sect. 3.3, 3.4).

3.1 *decapentaplegic*: A TGF-β Family Member

The *decapentaplegic* (*dpp*) gene was isolated by a chromosomal walk from an adjacent middle repetitive element (ST. JOHNSTON et al., in preparation). DNA sequence analysis of cDNA clones indicated an open reading frame that encoded a protein similar in structure and sequence to members of the TGF-β family of growth and differentiation factors (PADGETT et al. 1987). The structural similarities include an amino terminal hydrophobic sequence that could act as a signal sequence for secretion of the protein product and dibasic residues that could be sites of cleavage of a precursor into amino terminal and carboxy terminal polypeptides. Within the carboxy terminal 100 amino acids, the *dpp* sequence contains seven cysteine residues that are conserved in position in the aligned sequences of the TGF-β family, including those for TGF-β1, TGF-β2, inhibin B, inhibin A, Mullerian inhibiting substance and *veg*1 (PADGETT et al. 1987; WEEKS and MELTON 1987). The sequence of the *Drosophila* protein has between 27%—48% amino acid sequence identity to these other proteins within this 100 amino acid carboxy terminal domain. Biochemical analysis indicates that the *dpp* protein is secreted from *Drosophila* tissue culture cells and processed to an amino terminal and carboxy terminal portion (PANGANIBAN et al., in preparation).

As in the case of most of the other growth factor and proto-oncogene homologs, the *dpp* gene product acts at several different times during development (see Sect. 4.1). By analysis of RNAs from developmentally staged animals, there are three major transcripts of 3.3, 3.8, and 4.3 kb from the *dpp* gene (ST. JOHNSTON et al., in preparation). The 3.3-kb transcript is present in early embryos but is absent in embryos after 12 h of development. The 3.8-kb transcript is detected first during midembryogenesis (8–12 h) and is present at all later stages, althought at highest levels in 12- to 24-h embryos and pupae. The 4.3-kb transcript is detected in early embryos and all later stages. Several other lower abundance *dpp* transcripts have also been detected. The transcripts all encode the same protein but differ in their 5′ untranslated exons. Transcriptional initiation of the alternate 5′ exons is controlled by separate promoters that are distributed across 23 kb of genomic DNA. The function of this complex transcriptional regulation is currently under investigation. It has been demonstrated that one promoter, driving the 3.3-kb transcript, is sufficient for the epidermal dorsal/ventral functions of *dpp* (HOFFMANN and GOODMAN 1987).

Earliest expression of *dpp* is detected at syncytial blastoderm (nuclear cycle 11) on the dorsal surface of the embryo that will eventually give rise to the dorsal hypoderm and many of the highly specialized structures of the head and tail of the animal (ST. JOHNSTON and GELBART 1987). The level of the dorsally localized RNA increases during cellularization of the blastoderm. Cells extending around approximately 40% of the circumference of the blastoderm express the *dpp* transcript. As the cells from the dorsal surface migrate ventrally to produce the germ band, the dorsal band of *dpp* expression splits and migrates to yield two bands of cells, one along either side of the embryo. A few cells remain at the dorsal surface and form a squamous epithelium called the amnioserosa; *dpp* is not expressed in these cells. Throughout the process of germ band elongation, the cells of the presumptive dorsal epidermis contain *dpp* transcripts, but as germ band shortening begins, the pattern of *dpp* expression changes again, and the *dpp* transcript is now observed in four separate locations: two laterally symmetric bands of cells along the dorsalmost cells of epidermis that border the amnioserosa, and two laterally symmetric bands of cells close to the boundary between the dorsal epidermis and the ventral epidermal/neurogenic region. These cells contain *dpp* transcript during the process of germ band shortening and while dorsal closure takes place, but by the time dorsal closure is completed *dpp* is not detectable in the epidermis cells.

After dorsal closure, *dpp* mRNA is detected in the visceral mesoderm of the thoracic and abdominal regions. The visceral mesoderm forms the smooth muscle cells between the somatic mesoderm and the gut; it functions as the musculature of the gut. The two regions of the visceral mesoderm that express *dpp* continue to do so throughout the formation and differentiation of this structure. The cells that express *dpp* are at the boundaries of the precursor tissues that link the anterior midgut and the posterior midgut. Expression of *dpp* in the extreme anterior end of the animal is also associated with structures actively involved in migration during organogenesis, e.g., the cells of the stomodeal invagination and the cells of the pharynx that give rise to the stomatogastric nervous system. The *dpp* gene is also expressed later in the imaginal disks (see Sect. 4.1).

The multiple roles of *dpp* in embryonic development are indicated best by the number of structures affected by mutations in *dpp* (SPENCER et al. 1982; SEGAL and GELBART 1985; IRISH and GELBART 1987). The phenotypes of complete loss-of-function (null) mutations in *dpp* include defects in gastrulation and transformation of the dorsal larval hypoderm to ventral larval hypoderm. The cuticles formed by the mutant embryos are segmented, but the bands of denticle hairs, normally found on the ventral surface, extend around the entire dorsal/ventral circumference of the cuticle. In addition, structures at the anterior and posterior ends of the cuticle are missing or malformed. Deletions of the *dpp* protein-coding regions are haploinsufficient (*hin*) alleles $dpp^{hin}/+$ animals die as embryos. The dead embryos exhibit relatively normal dorsal/ventral polarity on the larval cuticle but severe defects of head and caudal structures; the dosage requirement for *dpp* in the development of these structures is not understood at this time. *dpp* is also critical for completion of organogenesis in late embryogenesis, particularly with respect to the formation of ectodermal structures associated with the continuity of the gut. Defects in gut formation, and perhaps in other internal organs, are the cause of the larval lethality of some *dpp* mutant alleles (SEGAL and GELBART 1985). In addition to these embryonic functions, *dpp* mutant alleles affect development of the imaginal disks (Sect. 4.1).

The *dpp* gene is one of several genes that are known to participate in the determination of dorsal/ventral pattern formation in the early embryo (ANDERSON 1987). Twelve of these genes are maternal-effect genes, expressed during oogenesis, and they include the *dl* gene discussed above. Several other genes, including *dpp*, *zen*, *twist*, and *snail*, are expressed in the early zygote. To understand the process of dorsal/ventral pattern formation, it will be important to know how these genes affect one another. Mutations in the maternal-effect dorsalizing gene *dl* lead to uniform expression of *dpp* throughout the cells of the blastoderm embryo, in contrast to the normal dorsally localized pattern of *dpp* expression (D. N. GUNN and F. M. HOFFMANN, unpublished observation). In turn, mutations in *dpp* have an effect on the expression of another dorsally localized gene, *zen*. Although the early blastoderm expression of *zen* is normal, later expression in the amnioserosa is not detected in *dpp* mutant embryos (C. RUSHLOW, personal communication). Thus, the *dpp* product, a growth factor homolog, may occupy a pivotal site in a regulatory circuit affected by the maternal dorsal/ventral product(s) and regulating other zygotic dorsal/ventral loci.

3.2 *wingless*: An *int-1* Homolog

The murine *int-1* proto-oncogene is the site of insertion of mouse mammary tumor virus (Nusse et al. 1984). Pertubation of *int-1* regulation by retroviral insertion is a critical step in the formation of retrovirally induced mammary tumors in the mouse. The mouse gene encodes a 370 amino acid protein that appears to be secreted from the cells to act as a growth factor with autocrine or paracrine activities (Papkoff et al. 1987). The mouse *int-1* DNA was used as a probe to identify a cross-hybridizing DNA sequence from *Drosophila* (Rijsewijk et al. 1987). The same *Drosophila* sequence was isolated independently by transposon tagging of the *wingless* (*wg*) gene (Baker 1987). Analysis of the sequence of cDNA and genomic clones indicated the presence of five coding exons aligned over 10 kb of genomic DNA. The open reading frame encodes a 468 amino acid polypeptide that includes an amino-terminal signal sequence and that, overall, has 54% amino acid identity with the mouse *int-1* protein.

A 3- to 3.2-kb *wg* transcript is detected on Northern blots in RNA samples from animals at all stages of development (Rijsewijk et al. 1987; Baker 1987). The messenger RNA is least abundant in preblastoderm embryos and most abundant in 3- to 6-h embryos (early gastrulation) and in pupae (Baker 1987). The *wg* RNA was localized in the embryo by in situ hybridization to embryos (Baker 1987, 1988c). At cellular blastoderm, *wg* transcripts are detected initially at the anterior tip (the cells of the presumptive foregut) and in a posterior ring (cells of the presumptive hindgut and anal structures). The cells of the dorsal head region are the next to express *wg*, followed by 14 stripes of cells, one stripe in each parasegmental region 0–13. The parasegments are repeated embryonic primordia along the anterior/posterior axis of the embryo indicated morphologically by grooves in the ectoderm and molecularly by patterns of gene expression (reviewed by Akam 1987). During germ band extension additional regions of *wg* expression are detected, including two more areas in the head and in parasegment 14. At this time a total of 21 domains of *wg* expression exist. Transcript is observed at the extreme anterior end in the primordial cells of the stomodeum, clypeolabrum, and procephalic lobes, and at the extreme posterior end in the cells of the analia and hindgut. Sixteen bands of *wg* RNA are detected in the epidermis after germ band shortening, as stripes three to five cells wide immediately anterior to the anterior/posterior compartment boundary of each segment (Baker 1987). The *wg* transcript is also expressed transiently in the precursor cells of the mesoderm and in the developing central nervous system. The segmental pattern of *wg* expression is similar to the pattern observed for several other *Drosophila* genes, including *engrailed* and *gooseberry*, that determine segmental polarity (reviewed by Ingham 1988).

In the absence of the *wg* gene product, the cells in each segment of the larval cuticle differentiate as if they were all in the anterior compartment of a segment and give rise to a larval cuticle with a continuous field of ventral denticle hairs and no segment boundaries; there is also a reduction in dorsal cuticle. Pole cell transplants show that there is no maternal requirement for *wg* gene product; temperature-sensitive mutations show that the embryonic requirement for *wg* function begins at gastrulation and continues until germ band retraction is complete (Baker 1988a). Analysis of the mutant phenotypes indicates that the absence of *wg* expression in a subset of the cells in each metameric segment affects the cell fate of adjacent cells, for example, cells in the posterior compartment of each segment, which do not express *wg* during their normal development. This implies that the *wg* product, either directly or indirectly, must act extracellularly to affect the determination or viability of adjacent cells. One example of the action of *wg* on the adjacent cells of the posterior compartment is its role in maintaining in these cells the level of expression of another segment polarity gene, *engrailed* (Martinez-Arinas et al. 1988). The regulatory interactions between cell autonomous functions, like the nuclear localized *engrailed* protein, and cell-nonautonomous functions, like the extracellular *wg* protein, result in patterns of gene expression that determine the different fates of cells within each segment. This process requires the action of specific DNA-binding proteins, like *engrailed*, to regulate the expression of the appropriate genes, but it also requires a mechanism for cells to communicate their impending cell fates in order that adjacent cells might respond appropriately. It is this process of communication that requires the *wg* protein and, potentially, other as yet undiscovered growth factor homologs in *Drosophila*.

3.3 *Notch* and *Delta*: Proteins with EGF-like Structures

Notch DNA was obtained by isolating a DNA clone of a chromosomal rearrangement that brings *Notch* sequences adjacent to a previously isolated DNA sequence (Artavanis-Tsakonas et al. 1983; Kidd et al. 1983). DNA sequencing of the *Notch* gene revealed an open reading frame of 2703 amino acids with a putative extracellular domain containing 36 repeated segments, each consisting of approximately 40 amino acids, with similarity to the EGF sequence (Wharton et al. 1985; Kidd et al. 1986). The location of the cysteine residues in each of the repeats is highly conserved. The extreme amino terminus of the protein contains a hydrophobic region that could function as a signal peptide; an internal 21 amino acid hydrophobic region, flanked by charged residues, could function as a membrane-spanning domain. The large number of EGF-like repeats is unique to the *Notch* protein in comparison with mammalian proteins. The mouse EGF precursor contains only nine repeats, one of which is cleaved out and secreted as mature EGF. Other proteins with EGF-like repeats, such as plasminogen activator, urokinase, TGF-α, and vaccinia virus 19 K protein, each contain only one EGF-like sequence.

Delta was cloned using a chromosomal walk from an adjacent P-element insertion (Vassin et al. 1987) and independently by a chromosomal walk across an inversion breakpoint that brought sequences from the bithorax complex adjacent to the sequences of the *Delta* gene (Alton et al. 1988; Kopczynski et al. 1988). Mutant lesions in the *Delta* gene were mapped on the DNA and indicated that a 35-kb region of DNA was critical for wild-type *Delta* function. The full-length protein is 832 amino acids in length and contains sequences appropriate for a signal peptide, an extracellular domain with nine EGF-like repeats, a membrane-spanning region, and an intracellular domain.

The expression of the *Notch* gene in the developing embryo has been examined by in situ hybridization (Hartley et al. 1987). The 10.5-kb RNA of the *Notch* gene is detected throughout embryogenesis, beginning with the presence of a maternally provided transcript in the precellular blastoderm. Zygotic transcription begins in the cellular blastoderm with no bias for the cells in the neurogenic region. At the time of gastrulation, the *Notch* transcript is detected in the cells of the presumptive mesoderm that invaginate through the ventral furrow. Later in embryogenesis, the *Notch* transcript is confined to the cells of the epidermis. During the development of the ventral nervous system, the level of *Notch* RNA decreases.

The transcriptional expression of the *Delta* gene is quite complex, consisting of two maternal transcripts of 4.5 and 3.6 kb, and four zygotic transcripts, two of 5.4 kb, one of 3.5 kb, and one of 2.8 kb (Kopczynski et al. 1988). These multiple transcripts result from alternate splicing and alternate sites of polyadenylation. The function of the maternal transcripts is not known, although the *Delta* transcripts expressed maternally encode the same protein product as the zygotic transcripts. The time of expression of the zygotic *Delta* transcripts is between 3 and 6 h after fertilization; this coincides with the time that *Delta* function is required for neurogenesis. At cellular blastoderm, the *Delta* transcripts are located in the cells that will become the neurogenic ectoderm, the anterior and posterior midgut, and the dorsal portion of the hindgut (Vassin et al. 1987). The higher levels of *Delta* transcript in the ventral neurogenic regions contrast with the more uniform distribution of the *Notch* transcript. This ventral localization is transient in that *Delta* RNA is detected around the dorsal/ventral circumference of the embryo before gastrulation begins. During the germ band extension phase of gastrulation, all of the ectodermal cells contain *Delta* transcripts.

The development of the central nervous system during embryogenesis begins when the cells in the ventral ectodermal region make a choice between becoming neuroblasts or dermatoblasts, the precursors of the epidermal structures. The segregation of the neuroblasts begins at about 4 h into embryonic development and continues for approximately 3 h. In the absence of the *Notch* or *Delta* gene products, approximately three times the normal number of cells choose the neural fate over a dermal fate, leading to a hypertrophy of the nervous system and lethality to the embryo. In addition to *Notch* and *Delta*, four other *Drosophila* genes have been identified that produce hypertrophy of the embryonic nervous system when mutated: *big brain*, *mastermind*, *Enhancer of split*, and *neuralised* (Lehmann et al. 1983). Transplantation of mutant cells indicates that the defects caused by mutations in all of these genes but *Enhancer of split* can be rescued by surrounding wild-type cells (Technau and Campos-Ortega 1987). This suggests that the products of the other five genes affect the developmental fates of adjacent cells and not the fate of the cell in which the gene is expressed. This cell nonautonomous behavior is consistent with the existence of extracellular EGF-like domains in the *Notch* and *Delta* proteins which, by analogy to other proteins with similar domains, are likely to be involved in protein-protein recognition mediating intercellular communication. The importance of proteins containing the EGF-like motif in instructing cell fates and behaviors is also supported by the discovery of two such proteins involved in cell fate determination in *C. elegans*, *lin-12* and *glp-1* (Greenwald 1985; Yochem et al. 1988; Austin and Kimble 1987; Priess et al. 1987). Analysis of mosaic clones of *Notch* mutant cells indicates that the *Notch* protein does not act over long distances (Hoppe and Greenspan 1986), but the *Notch* and *Delta* gene products may act on adjacent cells in the neurogenic ectoderm and influence those cells to pursue a dermal rather than a neural lineage. The expression of *Notch* RNA in a wide variety of cells in the embryo and the effect of specific *Notch* mutant alleles on the phenotype of the adult eye and wing (DIETRICH and CAMPOS-ORTEGA 1984) suggest that the *Notch* product may have additional roles besides affecting the decision of embryonic cells to be dermatoblasts versus neuroblasts.

Other *Drosophila* sequences that cross-hybridize to the EGF-like sequence of *Notch* have been identified by low stringency DNA cross-hybridization (KNUST et al., 1987; ROTHBERG et al. 1988). A sequence from polytene chromosome region 95F was isolated and sequenced, and the amino acid sequence similarly was found to be between 33% and 40% identical with the EGF-like repeats in *Notch* and *Delta* (KNUST et al. 1987). A separate isolate mapped to 52D and may be identical with the previously identified gene *slit*, a gene required for the proper formation of the embryonic ventral nervous system (ROTHBERG et al. 1988). DNA sequencing revealed seven EGF-like repeats in the 52D sequence. The 52D probe detected a 9.5 kb transcript which is first expressed in the presumptive ectodermal cells of the blastoderm and subsequently in the cells of the lateral ectoderm. The most intense hybridization was observed along the midline of the embryo upon differentiation of the midline neuroepithelium. These cells are important for guiding the axonal outgrowths to produce the characteristic network of longitudinal and commissural axon bundles. The 52D protein was detected with antibodies in the midline epithelium cells at the time of their differentiation from the ectoderm. The protein becomes associated with axonal membranes that track over the midline epithelium cells. The neural pattern of expression of the 52D sequence reflects the pattern of neural disruption effected by *slit* mutations. Mutations in *slit* lead to a collapse of the ventral nervous system as if the midline epithelium cells do not instruct correct axonal pathfinding.

3.4 *flb* and *abl* Embryonic Functions

The *Drosophila* EGF receptor homolog *flb*, whose function in oogenesis was described in Sect. 2.1, is also expressed in the early blastoderm embryo. At the time of cellular blastoderm and gastrulation, there is a precipitous increase in the level of *flb* RNA. The level of *flb* transcripts was measured relative to the embryonic α_1-tubulin transcript and found to constitute 0.1%–0.2% of the embryonic poly-(A)$^+$RNA (KAMMERMEYER and WADSWORTH 1987). *flb* RNA is detected throughout embryogenesis and through the larval and pupal stages. Both *flb* transcripts are expressed during these periods, but in the adult the smaller transcript predominates. *flb* RNA is present at the periphery of the cellular blastoderm embryo and uniformly throughout the embryo midway through embryogenesis (SCHEJTER et al. 1986; KAMMERMEYER and WADSWORTH 1987).

Mutant alleles of *flb* have been isolated in screens for recessive lethal mutations affecting embryogenesis (NÜSSLEIN-VOLHARD et al. 1984) and in screens for recessive lethals in the chromosomal region containing *flb* (PRICE et al. 1989; SCHEJTER and SHILO, 1989). Mutant alleles of *flb* were distinguished from non-*flb* lethal alleles in the region by biochemical assays of the amount of phosphorylated receptor immunoprecipitated from heterozygous mutant animals. In *flb*/+ animals the amount of protein is half of the level observed in extracts from +/+ wild-type animals. Some of the mutant alleles, identified by their lack of genetic complementation with the *flb* null alleles, produce *flb* protein that retains in vitro tyrosine kinase activity. These mutations may specifically affect ligand binding or ligand-mediated activation of the tyrosine kinase (B.-Z. SHILO, personal communication).

The absence of zygotic *flb* expression in the embryo results in embryonic lethality. Defects are seen at about 9 h into embryogenesis in the morphogenetic movements of gastrulation, in the cellular structures of the anterior portions of the embryo, and in the embryonic central nervous system. The embryo secretes a rudimentary, unpatterned larval cuticle, hence the name *faint little ball*. The association of the *flb* transcripts with mitotically active cells in the larvae (KAMMERMEYER and WADSWORTH 1987) suggests that *flb* may be required for cellular proliferation. In the embryo, however, most of the cells undergo only two or three rounds of cell division postblastoderm (HARTENSTEIN and CAMPOS-ORTEGA 1985). The severity of the embryonic defects is difficult to reconcile simply with a failure of embryonic postblastoderm mitosis unless there are also significant amounts of cell death. The *flb* protein, therefore, seems to be required for embryonic cell viability.

Like *flb*, *abl* affects embryogenesis through its maternal expression (Sect. 2.2) and its embryonic zygotic expression. The *abl* tyrosine kinase gene is expressed at midembryogenesis specifically in the cells of the developing ventral nervous system (GERTLER et al. 1989). As the germ band begins to retract, the neuroblast cells begin to differentiate and coalesce into the embryonic central nervous system (CAMPOS-ORTEGA and HARTENSTEIN 1985). The differentiation of these cells is marked by the appearance of neural-specific antigens on the cell surfaces. The *abl* protein can be detected with antibodies specific to it in the projections of the differentiating cells. The protein is initially present in the axons extending across the ventral midline in axon bundles called commissures. Shortly thereafter, the protein remain higher in the longitudinal bundles than in the commissure bundles, even though staining with antibodies to other neural antigens indicates that the number of axons is similar in the two types of axon bundles. This suggests that the *abl* protein is present in more of the cells that extend axons in the longitudinal bundles than cells extending axons in the commissure bundles. As embryogenesis proceeds, and the ventral nervous system begins to condense to form the ventral ganglion, *abl* protein is still detectable in the axons. We do not know whether the preferential localization of the protein in the axonal projections as opposed to the cell bodies themselves is important. The cells of the peripheral nervous system do not have detectable levels of the *abl* tyrosine kinase.

Mutations that eliminate zygotic *abl* expression do not produce gross defects in the ventral nervous system in the embryo; however, subtle changes in axonal projections and/or connections would not have been detected by the methods used. Mutations in *abl* generally lead to lethality during the pupal period (HENKEMEYER et al. 1987). The absence of a detectable mutant phenotype in the embryonic central nervous system may be due to the presence of residual *abl* protein from the maternally supplied *abl* transcript or because the absence of *abl* protein alone does not have major ramifications on the architecture of the central nervous system. The latter might be expected if the *abl* cytoplasmic tyrosine kinase modulates cellular processes involved in neural differentiation, but is not essential for those processes to occur. Our observations suggest, however, that *abl* mutations make the development of the embryonic central nervous system more sensitive to mutant lesions in other genes, suggesting that *abl*'s role in the development of the embryonic central nervous system is masked in a wild-type genetic background (GERTLER et al. 1989). The identification of the products encoded by these other genes may provide clues to the regulatory pathway modulated by the *abl* protein.

4 Imaginal Disk Development: Cell Proliferation and Specification of Pattern

During the three larval instars the animal grows in size, largely through the increase in cell size of the polyploid larval cells. An increase in cell number occurs only in those tissues with diploid cells, primarily in the nervous system and in the specialized epithelial sheets called imaginal disks. Cellular proliferation occurs in these tissues in preparation for the construction of the nervous system and the exoskeleton of the adult during metamorphosis. The cells of the 19 imaginal disks produce all of the adult cuticular structures except the abdominal cuticule. Cell proliferation is a highly regulated process in imaginal disk development (BRYANT 1987). In the wing disk, for example, 10–20 embryonic founder cells produce 50000 cells in the mature wing disk, i.e., approximately 11–12 cell generations. The cells in the imaginal disks are specified to differentiate specific cuticular structures depending largely on their positions within the developmental field of each imaginal disk. Specific adult structures, e.g., small distinct groups of bristles or sensory organs, have been fate-mapped to specific cells in the imaginal disk. The two processes, cellular proliferation and specification of cell fate, are tightly coupled as indicated in disk regeneration experiments. When a portion of an imaginal disk is removed, cell proliferation is stimulated and continues only until sufficient cells are produced to regenerate the portion that was removed, i.e., the regeneration of a complete developmental field leads to cessation of cell proliferation (BRYANT and SIMPSON 1984).

The proliferation and specification of the imaginal disk cells occur during the larval stages of development. During the next stage, the pupal period, metamorphosis of the larval to the adult form occurs. The imaginal disks evert their highly folded epithelial sheets, which fuse with one another and secrete the adult cuticle of the head, thorax, and genitalia. The abdominal cuticle is produced by cells derived from the abdominal histoblast nests that proliferate and differentiate during the pupal period.

Several of the genes introduced in earlier sections are required for proper development of the imaginal disks. *dpp*, *wg*, and *pole hole* are required for cell viability or proliferation in the imaginal disks (Sects. 4.1, 4.3). *abl* is required for neural differentiation in the eye imaginal disk (Sect. 4.3), as is the *sevenless* receptor tyrosine kinase (Sect. 4.2). While it is quite likely that many of these genes also have important functions during the pupal stage, none of the pupal functions have been well described and, thus, they are not discussed here.

4.1 Diffusible Factors in the Disk: *dpp* and *wg*

The expression of *dpp* RNA in the imaginal disks has been localized both by genetic criteria and by molecular probes to a subset of the cells in the imaginal disk along the anterior/posterior compartment boundary in the imaginal disks. Analysis of mosaic imaginal disks that contain both wild-type and *dpp* mutant cells indicates that the *dpp* product produced by a few cells along the anterior/posterior compartment boundary affects all of the cells in the disk, i.e., *dpp* functions in a cell nonautonomous manner consistent with its molecular similarity to a diffusible growth factor (Spencer 1984; Posakony 1987).

As discussed in Sect. 3.1, null mutations in *dpp* cause embryonic lethality; however, other mutations have been recovered at *dpp* that retain the embryonic functions of *dpp* but exhibit mutant phenotypes in the adult structures derived from the imaginal disks (Spencer et al. 1982). This is analogous to the specific *flb* alleles described in Sect. 2.1 that retain *flb* embryonic functions but exhibit a female sterile phenotype. As discussed below, a similar set of disk-specific mutant alleles has been recovered in the *wg* gene. The *dpp* mutations that specifically affect imaginal disk development all disrupt *dpp* DNA sequences in a 20-kb 3′ non-transcribed region; none of these disk-specific mutations disrupts the *dpp* coding sequences (Blackman et al. 1987; St. Johnston et al., in preparation). This 3′ non-transcribed regulatory region contains several *cis* regulatory elements that behave like cell-specific transcriptional enhancers (R. K. Blackman and W. M. Gelbart, personal communication; J. D. Masucci and F. M. Hoffmann, unpublished observations).

The mutant phenotypes caused by the *dpp* disk-specific mutations suggest that *dpp* is required in the imaginal disks for cell proliferation (Spencer et al. 1982; Bryant 1988). Disk-specific mutations in *dpp* result in the absence of structures from the distal portions of all imaginal disk-derived structures of the adult cuticle. Chromosomal rearrangement breakpoints that fall furthest from the coding sequence, on the order of 20 kb, cause relatively mild defects in the wing blade formed from the wing imaginal disk. Chromosomal rearrangements that remove the 3′ *cis* regulatory elements 10 kb away from the coding sequence cause defects in all of the adult structures derived from imaginal disks. The defects are severe: for example, virtually the entire wing blade is missing in this class of mutations. Lesions falling within 5 kb of the 3′ end of the *dpp* coding sequence lead to lethality at the pupal stage owing to the absence of sufficient imaginal cells to form an intact adult cuticle during metamorphosis.

The absence of *dpp* function in the imaginal disks produces cell death in the disks and leads to the formation of clusters of condensed material. The amount of cell death is greater in regions of the imaginal disk that give rise to the distal structures of the adult cuticle. This might be explained by the cells having different quantitative requirements for the *dpp* product. The defects caused by these mutations have been examined by transplantation of mutant imaginal disks (BRYANT 1988). Isolated *dpp* mutant imaginal disks transplanted into wild-type hosts were induced to undergo metamorphosis and the cuticular structures formed by the implanted disks examined. Disks derived from *dpp* mutant animals in which virtually all of the 3′ *cis* regulatory elements are mutationally removed produced structures only from the anterior notum and proximal anterior wing margin. If the mutant disks were cultured in female abdomens for 7 days to permit cell proliferation before being induced to metamorphose, the additional time had no effect in alleviating the loss of structures produced from the disk. Fragments of wild-type disks cultured in the same way will undergo cellular proliferation during the culture period, and these additional cells will produce structures that will either duplicate the structures to be formed by the implanted fragment or will regenerate structures that are absent from the implanted fragment. By mixing wild-type tissue with *dpp* tissue and allowing time for cellular proliferation by culture in a female abdomen, the *dpp* tissue was induced to produce many more structures during metamorphosis (BRYANT 1988). This finding is consistent with the production of a diffusible factor from the wild-type tissue that can compensate for the defect in the *dpp* mutant tissue.

In third instar larvae, *wg* RNA is detected in the imaginal disks but not in the larval brain or the polyploid larval cells (BAKER 1988b). *wg* RNA is detected in the wing, haltere, leg, antennal, and eye disks, but only a portion of the cells in these disks express the *wg* gene. In the wing and haltere disks, the cells expressing *wg* form a band across the wing pouch and a ring around the wing pouch. Some of the cells that will give rise to the notum also express *wg*. In the leg disks, the cells that express *wg* occupy a wedge-shaped area that will give rise to structures on the ventral anterior surface of the leg.

The requirement for *wg* function in imaginal disks is indicated by the loss of distal wing blade and haltere structures and concurrent duplication of proximal thoracic structures in flies carrying the wg^1 allele. The effect of the wg^1 mutations is to change the fate of cells in the wing and haltere imaginal disks resulting in duplications of notum structures on the adult cuticle. This effect is a homeotic transformation of wing blade to notum cell fate in contrast to the simple loss of structures observed in *dpp* mutations affecting imaginal disks. Generation of wg^1 cell clones in an otherwise wild-type background was used to demonstrate that *wg* function, like *dpp*, is cell nonautonomous (MORATA and LAWRENCE 1977). The nonautonomy of *wg* function has been confirmed in mitotic recombination experiments with the stronger wg^L allele (BAKER 1988a). The wg^L mutant clones arose at a frequency and size comparable to the wild-type control clones and always gave rise to wild-type structures, indicating that the loss of function of *wg* in a particular group of cells could be rescued by the wg^+ function present in surrounding cells. The nonautonomy indicates that the *wg* protein product may be secreted from cells and act like a diffusible growth factor.

Although the wg^1 mutation affects only the wings, halteres, and eye shape, the wg^{cx3} allele affects other imaginal disk-derived structures. These mutant animals die in the pupal case, as do *dpp* mutant animals with severe disk abnormalities. *wg* mutant animals dissected from pupal cases show wing and haltere transformations to notum, absence of antennal structures, duplications of leg structures, and defects in the labrum, proboscis, and abdominal ventral sternites (BAKER 1988b). Temperature-sensitive *wg* mutations have also been used to address the requirement for *wg* in the imaginal disks. Shifting to the nonpermissive temperature during the first larval instar produces a wide range of defects in the adult cuticle of the pharate adults, indicating a requirement for *wg* function in the imaginal disks and the abdominal histoblasts (BAKER 1988a). The *wg* mutant alleles that provide normal *wg* embryonic functions but defects in imaginal disk development all map 3′ of the *wg* transcriptional unit (BAKER 1987). The mutations may affect 3′ *cis* regulatory elements required specifically for disk expression of *wg*, analogous to the 3′ *cis* regulatory sequences required for imaginal disk expression of the *dpp* gene.

4.2 *sevenless*: A Receptor Tyrosine Kinase

DNA from the *sevenless (sev)* gene was isolated by microdissection and cloning of the polytene chromosome region 10A1,2 and, independently, by transposon tagging (HAFEN et al. 1987; BANERJEE et al. 1987a). DNA sequences sufficient for *sev* function were identified when a 15-kb DNA genomic fragment, introduced into the genome by P-element-mediated germline transformation, compensated for the lack of *sev* gene function and provided wild-type eye development. The sequence of 2554 amino acids predicted from the DNA sequence of a *sev* cDNA indicates that the *sev* protein has structural similarities to the class of growth factor receptor tyrosine kinases (BASLER and HAFEN 1988a). The carboxy terminal domain contains a region with 58% amino acid identity to the tyrosine kinase domain of the c-*ros* proto-oncogene (BASLER and HAFEN 1988b). Elimination of *sev* tyrosine kinase activity by changing a single amino acid, lysine 2242, at the putative ATP binding site eliminates the ability of the *sev* protein to rescue the *sev* mutant phenotype, i.e., tyrosine kinase activity is required for *sev* function. The *sev* protein contains two regions, of 22 and 24 amino acids respectively, with sequence characteristics of membrane-spanning regions. The *sev* protein is processed to an amino-terminal subunit of 220 kd and a carboxy-terminal subunit of 60 kd (E. HAFEN, personal communication). The two subunits both span the membrane, thereby explaining the presence of two transmembrane domains in the primary sequence. The proteolytic processing of the *sev* protein into subunits distinguishes it from other receptor tyrosine kinases, e.g., the EGF receptor and platelet-derived growth factor (PDGF) receptor tyrosine kinases are single polypeptides.

The development of the *Drosophila* compound eye has been described in detail (TOMLINSON 1985). The 700 ommitidia that make up the adult eye differentiate from the eye imaginal disk epithelium during the third larval instar. Each ommitidium consists of a defined set of a few specific cell types: eight photoreceptor cells, four cone cells, three types of pigment cells, a sensory hair and nerve cell. The cells differentiate in a specific order and geometric arrangement; the fate of each cell depends on its position within the cluster and the specific cell contacts made (TOMLINSON and READY 1987). The eight retinal cells form a regular trapezoidal pattern with cells 1–6 (R1–R6) on the perimeter and cells 7 and 8 (R7, R8) occupying the center position. Mutant alleles of *sev* cause the cell that normally would give rise to R7, the last of the eight retinal cells to differentiate, to follow a different cell fate and become a cone cell. Each of the ommitidia develops with only seven retinal cells. The *sev* gene must be expressed in the cell for it to become R7, i.e., the *sev* gene product acts in a cell autonomous fashion.

The 8.2-kb RNA transcript of *sev* is present in adult heads but not in the remainder of the fly (HAFEN et al. 1987). The *sev* mRNA is detected in eye imaginal disks by in situ hybridization and is first observed in the morphogenetic furrow, the morphological manifestation of a wave of coordinated cell division and movement that traverses the eye disk and that begins the differentiation of the uniform eye disk epithelium into the specialized cells of the ommatidia. Antibodies made against synthetic peptides representing *sev* sequences were used to localize the protein to the developing retinal cells in the eye disk (TOMLINSON et al. 1987; BANERJEE et al. 1987b). Electron microscopic examination of the cellular localization of the *sev* protein indicated that the protein was first detectable in two differentiating retinal cells, R3 and R4, and two cells of unknown fate (TOMLINSON et al. 1987). After the last division of the cells takes place, most of the cells surrounding the cluster express the *sev* protein (BANERJEE et al. 1987b). As cells complete their differentiation, the level of protein in R1 and R6 decreases, whereas the level of protein in R7 and the anterior and posterior cone cells increases. The highest concentration of protein occurs at the apical surfaces of the cells in a region of highly specialized cell-to-cell contacts involving adherens-type junctions. The *sev* protein present in cells R3, R4, and R7 is localized to the portions of the cell membrane that are in contact with cell R8. The specification of a particular cell as R7 is thought to occur because of the specific cell contacts made by the precursor cell (TOMLINSON and READY 1987; TOMLINSON et al., 1987). The cell that becomes R7 is the only cell in the cluster that makes contact with R1, R6, and R8. The function of the *sev* protein may be to recognize a signal molecule on the adjacent cells, through cell-to-cell contacts, and to activate the *sev* tyrosine kinase activity. The intracellular action of the *sev* tyrosine kinase, integrated with other regulatory information, would instruct the cell to differentiate to R7.

4.3 Other Protein Kinases in the Disk: *flb*, *pole hole*, and *abl*

Tissue localization of the *flb* transcript was determined by in situ hybridization to sections of developmentally staged organisms. *flb* RNA is not observed in polyploid cells of the larva, but the RNA is present in the mitotically active imaginal disks and neural cells in the brain and thoracic ganglia (SCHEJTER et al. 1986). The RNA in the larval brain is located in the mitotically active subregions of the brain cortex (KAMMERMEYER and WADSWORTH 1987). With the exception of the *top* alleles, most *flb* mutant alleles cause embryonic lethality; therefore, to study the effect of the *flb* mutations on the development of the imaginal disks or larval nervous system, mitotic clones of *flb* mutant cells will need to be induced in a *flb*/+ heterozygous animal or a temperature-sensitive *flb* allele isolated.

The c-*raf* homolog *pole hole* is required for proliferation of the imaginal disk and neural cells. *pole hole* RNA is detected throughout development (MARK et al. 1987; NISHIDA et al. 1988). As discussed in Sect. 2.4, the embryonic functions of *pole hole* are effectively supplied by the *pole hole* maternal mRNA. The zygotic mutant phenotype caused by the *pole hole* mutations is lethality at the larval or pupal stages.

pole hole mutant larvae are missing cells specifically in the tissues that proliferate during the larval stages: the imaginal disks, the brain lobes, and the lymph glands (PERRIMON et al. 1985; NISHIDA et al. 1988).

Expression of the *abl* tyrosine kinase has been detected specifically in the eye imaginal disk at the time of retinal cell differentiation, about the same time that the *sevenless* gene is expressed (R. L. BENNETT and F. M. HOFFMANN, in preparation). The *abl* protein is detected in all of the retinal cells, in the cell bodies, and in the axons extending into the optic lamina. Mutations in *abl* disrupt the normal number and pattern of retinal cells producing a disorganized array of differentiated retinal cells (HENKEMEYER et al. 1987). The effect on retinal cell determination and the expression of *abl* in the differentiating embryonic central nervous system (Sect. 3.4) suggest that this cytoplasmic tyrosine kinase has a specific function in neural cell development and does not reflect a more general vital metabolic or cell proliferation function. In addition to its expression in the eye imaginal disk, developmental Northern analysis reveals a transient increase in *abl* RNA levels in the metamorphosing pupae (TELFORD et al. 1985; K. D. RAND and F. M. HOFFMANN, unpublished observations). The pupal expression has not been localized to a tissue, but the time of expression coincides with the differentiation of the adult central nervous system (TRUMAN and BATE 1988). Mutations in *abl* result in pupal lethality, and the few animals that eclose from the pupal case are weak, uncoordinated, and die prematurely (HENKEMEYER et al. 1987). *abl* may be required for proper differentiation of the adult nervous system during pupation, consistent with its expression in the differentiating embryonic central nervous system and the developing retinal cells of the eye.

5 Future Directions

It is apparent from the work reviewed here on nine *Drosophila* genes that *Drosophila* homologs of vertebrate proto-oncogenes, growth factors, and growth factor receptors can participate in essential processes throughout development. Two of the genes, *dl* and *sev*, are required at single times during development, but the products of the other seven genes participate in regulatory pathways at several different times during development. There is much to be learned from additional study of these nine genes, but it will also be important to expand the number of *Drosophila* homologs under investigation to obtain a better perspective of the range of developmental processes affected by these genes. Several genes have been identified that are not discussed in detail here because mutant alleles are not yet available. For example, three *ras* homologs have been reported (NEUMAN-SIBERBERG et al. 1984; SCHEJTER and SHILO 1985; MOZER et al. 1985; BROOK 1987), and although no loss-of-function mutations have been reported, expression of a *Drosophila ras* protein mutated at amino acid 14, homologous to position 12 of vertebrate *ras*, results in a variety of phenotypic defects including disruption of eye development (BISHOP and CORCES 1988). Two *src* homologous genes exist in *Drosophila* (SIMON et al. 1983, 1985; GREGORY et al. 1987; VINCENT et al., in press). Both are expressed during oogenesis as part of the maternal contribution to early embryogenesis. Expression is also detected in differentiating neural tissues as well as other sites. *Drosophila* sequences also have been identified that are similar to the *myb* and *ets* proto-oncogenes (KATZEN et al. 1985; PRIBYL et al. 1988), the insulin receptor (PETRUZZELLI et al. 1986), and protein kinase C (ROSENTHAL et al. 1987). Mutant alleles in these genes will be needed to address their respective roles in *Drosophila* development.

The study of proto-oncogene, growth factor, and growth factor receptor homologs in *Drosophila* is providing information about normal *Drosophila* development and permitting the application of genetic strategies to study the roles of these proteins in normal cellular processes. The potential for using genetic strategies to identify new components of the regulatory pathways is just being tapped, but there are preliminary indications of regulatory interactions between the few genes identified thus far. For example, the *dl* protein may negatively regulate *dpp* expression, and the *dpp* protein may in turn positively regulate expression from one of the two *flb* promoters (PANGANIBAN et al., in preparation). The strong phenotypic interactions of *Notch* and *Delta* mutant alleles with the *Drosophila* gene *Enhancer of split* (SHEPARD et al., 1989) stimulated efforts to learn more about the product of *Enhancer of split*, which led to the finding that *Enhancer of split* encodes a protein homologous to the mammalian G proteins (HARTLEY et al., 1988). Application of similar genetic strategies to screen for genetic suppressors or enhancers of phenotypes caused by mutations in *abl*, *flb*, or *sev*, for example, will focus future efforts on new *Drosophila* genes that participate in the essential regulatory pathways upon which the products of these genes function. The potential for discovering new genes that interact with *Drosophila* proto-oncogenes, growth factors, and growth factor receptors is a primary reason for studying these kinds of genes in *Drosophila*.

Will the identification of new regulatory molecules involved in the control of cell proliferation or determination in *Drosophila* be useful for understanding these processes in mammalian cells? While the role of a particular homolog in the development of *Drosophila* may not be identical with its role in vertebrate development, the biochemical interactions of the protein product encoded by the homologous gene are likely to be well conserved. A *Drosophila* gene identified by its functional interaction with a *Drosophila* proto-oncogene homolog, for example, also may be highly conserved and have a homolog in mammalian cells. The mammalian gene could be identified with probes derived from the *Drosophila* gene, an approach that has been used very successfully to identify mammalian genes homologous to the *Drosophila* homeobox genes (HOLLAND and HOGAN 1988). Finally, it seems likely that a better understanding in any organism of such basic regulatory processes as the control of cell proliferation and determination will have positive ramifications on our understanding of these processes in the human.

6 References

Akam M (1987) The molecular basis for metameric pattern in the *Drosophila* embryo. Development 101: 1–22

Alton AK, Fechtel K, Terry AL, Meikle SB, Muskavitch MAT (1988) Cytogenetic definition and morphogenetic analysis of *Delta*, a gene affecting neurogenesis in *Drosophila melanogaster*. Genetics 118: 235–245

Anderson KV (1987) Dorsal-ventral embryonic pattern genes of *Drosophila*. Trends Genet 3: 91–97

Anderson KV, Nüsslein-Volhard C (1984) Genetic analysis of the dorsal/ventral embryonic pattern in *Drosophila*. In: Malacinski GM, Bryant SV (eds) Primers in developmental biology. Macmillan, New York pp 269–289

Anderson KV, Nüsslein-Volhard C (1984) Information for the dorsal-ventral pattern of the *Drosophila* embryo is stored as maternal mRNA. Nature 311: 223–227

Artavanis-Tsakonas S, Muskavitch MAT, Yedvobnick B (1983) Molecular cloning of *Notch*, a locus affecting neurogenesis in *Drosophila melanogaster*. Proc Natl Acad Sci USA 80: 1977–1981

Austin J, Kimble J (1987) *glp-1* is required in the germ line for regulation of the decision between mitosis and meiosis in *C. elegans*. Cell 51: 589–599

Baker NE (1987) Molecular cloning of sequences from *wingless*, a segment polyrity gene in *Drosophila*: the spatial distribution of a transcript in embryos. EMBO J 6: 1765–1773

Baker NE (1988a) Embryonic and imaginal requirements for *wingless*, a segment-polarity gene in *Drosophila*. Dev Biol 125: 96–108

Baker NE (1988b) Transcription of the segment-polarity gene *wingless* in the imaginal discs of *Drosophila*, and the phenotype of a pupal-lethal *wg* mutation. Development 102: 489–497

Baker NE (1988c) Localization of transcripts from the *wingless* gene in whole *Drosophila* embryos. Development 103: 289–298

Banerjee U, Renfranz PJ, Pollack JA, Benzer S (1987a) Molecular characterization and expression of *sevenless*, a gene involved in neuronal pattern formation in the *Drosophila* eye. Cell 49: 281–291

Banerjee U, Renfranz PJ, Hinton DR, Rabin BA, Benzer S (1987b) The *sevenless*$^{+}$ protein is expressed apically in cell membranes of developing *Drosophila* retina; it is not restricted to cell R7. Cell 51: 151–158

Basler K, Hafen E (1988a) Control of photoreceptor cell fate by the *sevenless* protein requires a functional tyrosine kinase domain. Cell 54: 299–311

Basler K, Hafen E (1988b) *Sevenless* and *Drosophila* eye development: a tyrosine kinase controls cell fate. Trends Genet 4: 74–79

Bennett RL, Hoffmann FM (in preparation) Expression of the *Drosophila* Abelson tyrosine kinase is associated with differentiation of neuronal cells

Bishop JG III, Corces VG (1988) Expression of an activated *ras* gene causes developmental abnormalities in transgenic *Drosophila melanogaster*. Genes Dev 2: 567–577

Blackman RK, Gramaila R, Koehler MMD, Gelbart WM (1987) Mobilization of hobo elements residing within the *decapentaplegic* gene complex: suggestion of a new hybrid dysgenesis system in *Drosophila melanogaster*. Cell 49: 497–505

Brock HW (1987) Sequence and genomic structure of *ras* homologues Dm*ras*85D and Dm*ras*64B and *Drosophila melanogaster*. Gene 51: 129–137

Bryant PJ (1987) Experimental and genetic analysis of growth and cell proliferation in *Drosophila* imaginal disks. In: Loomis, WF (ed) Genetic regulation of development: 45th Annual Symposium of the Society for Developmental Biology. AR Liss, New York pp 339–372.

Bryant PJ (1988) Localized cell death caused by mutations in a *Drosophila* gene coding for a transforming growth factor-β homolog. Dev Biol 128: 386–395

Bryant PJ, Simpson P (1984) Intrinsic and extrinsic control of growth in developing organs. Q Rev Biol 59: 387–415

Campos-Ortega JA, Hartenstein V (1985) The Embryonic Development of *Drosophila melanogaster*. Springer, Berlin Heidelberg New York, pp 1–84, 144–158

Delotto R, Spierer P (1986) A gene required for the specification of dorsal-ventral pattern in *Drosophila* appears to encode a serine protease. Nature 323: 688–692

Dietrich U, Campos-Ortega JA (1984) The expression of neurogenic loci in imaginal epidermal cells of *Drosophila melanogaster*. J Neurogenet 1: 315–332

Downward J, Yarden Y, Mayes E, Scrace G, Totty N, Stockwell P, Ullrich A, Schlessinger J, Waterfield MD (1984) Close similarity of epidermal growth factor receptor and v-*erb-B* oncogene protein sequences. Nature 307: 521–527

Durica DS, Restrepo MA, Thomas TL, Beckingham K (1987) Isolation and characterization of *abl* gene sequences in *Calliphora erythrocephala*. Gene 59: 63–76

Fukui M, Yamamoto T, Kawai S, Maruo K, Toyoshima K (1985) Detection of a *raf*-related and two other transforming DNA sequences in human tumors maintained in nude mice. Proc Natl Acad Sci USA 82: 5954–5958

Gertler FB, Bennett RL, Clark MJ, Hoffmann FM (1989) Drosophila *abl* tyrosine kinase in embryonic CNS axons: a role in axonogenesis is revealed through dosage-sensitive interactions with a new gene, *disabled*. Cell (In press)

Gilmore TD, Temin HM (1988) v-*rel* oncoproteins in the nucleus and in the cytoplasm transform chicken spleen cells. J Virol 62: 703–714

Goddard JM, Weiland JJ, Capecchi MR (1986) Isolation and characterization of *Caenorhabditis elegans* DNA sequences homologous to the v-*abl* oncogene. Proc Natl Acad Sci USA 83: 2172–2176

Greenwald I (1985) *lin-12*, a nematode homeotic gene, is homologous to a set of mammalian proteins that includes epidermal growth factor. Cell 43: 583–590

Gregory RJ, Kammermeyer KL, Vincent III WS, and Wadsworth SG (1987) Primary sequence and developmental expression of a novel *Drosophila melanogaster src* gene. Nol Cell Biol 7: 2119–2127

Haenlin M, Roos C, Cassab A, Mohier E (1987) Oocyte-specific transcription of *fs(1)K10*: *Drosophila* gene affecting dorsal-ventral developmental polarity. EMBO J 6: 801–807

Hafen E, Basler K, Edstroem J-E, Rubin GM (1987) *Sevenless*, a cell-specific homeotic gene of *Drosophila*, encodes a putative transmembrane receptor with a tyrosine kinase domain. Science 236: 55–63

Hartenstein V, Campos-Ortega JA (1985) Fate-mapping in wild-type *Drosophila melanogaster*. I. The spatio-temporal pattern of embryonic cell divisions. Roux's Arch Dev Biol 194: 181–195

Hartley DA, Xu T, Artavanis-Tsakonas S (1987) The embryonic expression of the *Notch* locus of *Drosophila melanogaster* and the implications of point mutations in the extracellular EGF-like domain of the predicted protein. EMBO J 6: 3407–3417

Hartley DA, Preiss A, Artavanis-Tsakonas S (in press) A deduced gene product from the *Drosophila* neurogenic locus, *Enhancer of split*, shows homology to mammalian G-protein beta subunit. Cell 55: 785–795

Henkemeyer MJ, Gertler FB, Goodman W, Hoffmann FM (1987) The *Drosophila* Abelson proto-oncogene homolog: identification of mutant alleles that have pleiotropic effects late in development. Cell 51: 821–828

Henkemeyer MJ, Bennett RL, Gertler FB, Hoffmann FM (1988) DNA sequence, structure, and tyrosine kinase activity of the *Drosophila melanogaster* Abelson proto-oncogene homolog. Mol Cell Biol 8: 843–853

Hoffman-Falk H, Einat P, Shilo B-Z, Hoffmann FM (1983) *Drosophila melanogaster* DNA clones homologous to vertebrate oncogenes: evidence for a common ancestor to the *src* and *abl* cellular genes. Cell 32: 589–598

Hoffmann FM, Goodman W (1987) Identification in transgenic animals of *Drosophila decapentaplegic* sequences required for embryonic dorsal pattern formation. Genes Dev 1: 615–625

Hoffmann FM, Fresco LD, Hoffman-Falk H, Shilo B-Z (1983) Nucleotide sequences of the *Drosophila src* and *abl* homologs: conservation and variability in the *src* family oncogenes. Cell 35: 393–401

Hoffmann FM, Henkemeyer MJ, Bennett RL, Gertler FB (1989) The Abelson tyrosine kinase in *Drosophila*: application of genetic strategies to understand its normal functions. In: Arlinghaus RB, Deisseroth A (eds) Molecular approaches to the study and treatment of Philadelphia chromosome positive leukemia. Marcel Dekker, New York, pp unknown

Holland PWH, Hogan BLM (1988) Expression of homebox genes during mouse development: a review. Genes Dev 2: 773–782

Hoppe PE, Greenspan RJ (1986) Local function of the *Notch* gene for embryonic ectodermal pathway choice in *Drosophila*. Cell 46: 773–783

Ingham PW (1988) The molecular genetic of embryonic pattern formation in *Drosophila*. Nature 335: 25–34

Irish V (1987) Cracking the *Drosophila* egg. Trends Genet 11: 303–304

Irish VF, Gelbart WM (1987) The *decapentaplegic* gene is required for dorsal/ventral patterning of the *Drosophila* embryo. Genes Dev 1: 868–879

Jansen HW, Patschinsky T, Bister K (1983) Avian oncovirus MH2: molecular cloning of proviral DNA and structural analysis of viral RNA and protein. J Virol 48: 61–73

Jurgens G, Wieschaus E, Nusslein-Volhard C, Kluding H (1984) Mutations affecting the pattern of the larval cuticle in *Drosophila melanogaster*: zygotic loci on the third chromosome. Roux's Arch Dev Biol 193: 283–295

Kammermeyer KL, Wadsworth SC (1987) Expression of *Drosophila* epidermal growth factor receptor homologue in mitotic cell populations. Development 100: 201–210

Katzen AL, Kornberg TB, Bishop JM (1985) Isolation of the proto-oncogene c-*myb* from *D. melanogaster*. Cell 41: 449–456

Kidd S, Lockett TJ, Young MW (1983) The *Notch* locus of *Drosophila melanogaster*. Cell 34: 421–433

Kidd S, Kelley MR, Young MW (1986) Sequence of the *Notch* locus of *Drosophila melanogater*: relationship of the encoded protein to mammalian clotting and growth factors. Mol Cell Biol 6: 3094–3108

Knust E, Dietrich V, Tepass V, Bremer KA, Weigel D, Vässin H, Campos-Ortega JÀ (1987) EGF homologous sequences encoded in the genome of *Drosophila melanogaster*, and their relation to neurogenic genes. EMBO J 6: 761–766

Kopczynski CC, Alton AK, Fechtel K, Kooh PJ, Muskavitch MAT (1988) *Delta*, a *Drosophila* neurogenic gene, is transcriptionally complex and encodes a protein related to blood coagulation factors and epidermal growth factor of vertebrates. Genes Dev Z: 1723–1735

Lehmann R, Jimenez F, Dietrich U, Campos-Ortega JA (1983) On the phenotype and development of mutants of early neurogenesis in *Drosophila melanogaster*. Roux's Arch Dev Biol 192: 62–74

Lev Z, Leibovitz N, Segev O, Shilo B-Z (1984) Expression of the *src* and *abl* cellular oncogenes during development of *Drosophila melanogaster*. Mol Cell Biol 4: 982–984

Lev Z, Shilo B-Z, Kimchie Z (1985) Developmental changes in expression of the *Drosophila melanogaster* epidermal growth factor receptor gene. Dev Biol 110: 499–502

Livneh E, Glazer L, Segel D, Schlessinger J, Shilo B-Z (1985) The *Drosophila* EGF receptor gene homolog: conservation of both hormone binding and kinase domain. Cell 40: 599–607

Mahowald AP, Kambysellis MP (1980) Oogenesis. In: Ashburner M, Wright TRF (eds) The genetics and biology of *Drosophila*, vol 2D. Academic, New York, pp 141–225

Mark GE, Seeley TW, Shows TB, Mountz JD (1986) *pks*, a *raf*-related sequence in humans. Proc Natl Acad Sci USA 83: 6312–6316

Mark GE, MacIntyre RJ, Digan ME, Ambrosio L, Perrimon N (1987) *Drosophila melanogaster* homologs of the *raf* oncogene. Mol Cell Biol 7: 2134–2140

Martinez-Arias A, Baker NE, Ingham PW (1988) Role of segment polarity genes in the definition and maintenance of cell states in the *Drosophila* embryo. Development 103: 157–170

Mayer BJ, Hamaguchi M, Hanafusa H (1988) A novel viral oncogene with structural similarity to phospholipase C. Nature 332: 272–275

Moelling K, Heimann B, Beimling P, Rapp UR, Sander T (1984) Serine- and threonine-specific protein kinase activities of purified gag-mil and gag-raf proteins. Nature 312: 558–560

Morata G, Lawrence PA (1977) The development of *wingless*, a homeotic mutation of *Drosophila*. Dev Biol 56: 227–240

Mozer B, Marlor R, Parkhurst S, Corces V (1985) Characterization and developmental expression of a *Drosophila ras* oncogene. Mol Cell Biol 5: 885–889

Neuman-Silberberg FS, Schejter E, Hoffmann FM, Shilo B-Z (1984) The *Drosophila ras* oncogenes: structure and nucleotide sequence. Cell 37: 1027–1033

Nishida Y, Hata M, Ayaki T, Ryo H, Yamagata M, Shimizu K, Nishizuka Y (1988) Proliferation of both somatic and germ cells is affected in the *Drosophila* mutants of *raf* proto-oncogene. EMBO J 7: 775–781

Nusse R, van Ooyen A, Cox D, Fung YKT, Varmus HE (1984) Mode of proviral activation of a putative mammary oncogene (*int*-1) on mouse chromosomes 15. Nature 307: 131–136

Nüsslein-Volhard C (1979) Maternal effect mutations that alter the spatial coordinates of the embryo of *Drosophila melanogaster*. In: Subtelney S, Koenigsberg IR (eds) Determination of spatial organization. Academic Press, New York, pp 185–211

Nüsslein-Volhard C, Frohnhöfer HG, Lehmann R (1987) Determination of anteroposterior polarity in *Drosophila*. Science 238: 1675–1681

Nüsslein-Volhard C, Wieschaus E, Kluding H (1984) Mutations affecting the pattern of the larval cuticle in *Drosophila melanogaster*. I. Zygotic loci on the second chromosome. Roux's Arch. Dev Biol 193: 267–282

Padgett RW, St Johnston RD, Gelbart WM (1987) A transcript from a *Drosophila* pattern gene predicts a protein homologous to the transforming growth factor-β family. Nature 325: 81–84

Panganiban GF, Neitzel M, Hoffmann FM (in preparation) Biochemical characterization of the *Drosophila* decapentaplegic protein, a member of the TGF-β family.

Papkoff J, Brown AT, Varmus H (1987) The *int*-1 proto-oncogene products are glycoproteins that appear to enter the secretory pathway. Mol Cell Biol 7: 3978–3984

Parks S, Spradling A (1987) Spatially regulated expression of chorion genes during *Drosophila* development. Genes Dev 1: 497–509

Perrimon N, Gans M (1983) Clonal analysis of the tissue specificity of recessive female-sterile mutations of *Drosophila melanogaster* using a dominant female-sterile mutation *Fs(1)*K1237. Dev Biol 100: 365–373

Perrimon N, Engstrom L, Mahowald AP (1985) A pupal lethal mutation with a paternally influenced maternal effect on embryonic development in *Drosophila melanogaster*. Dev Biol 110: 480–491

Petruzelli L, Herrera R, Arenas-Garcia R, Fernandez R, Birnbaum MJ, and Rosen OM (1986) Isolation of a *Drosophila* genomic sequence homologous to the kinase domain of the human insulin receptor and detection of the phosphorylated *Drosophila* receptor with an anti-peptide antibody. Proc Natl Acad Sci USA 83: 4710–4714

Posakony LM (1987) The role of the DPP-C in the development of the imaginal discs in *Drosophila melanogaster*. PhD dissertation, Harvard University

Pribyl LJ, Watson DK, McWilliams MJ, Ascione R, Papas TS (1988) The *Drosophila ets*-2 gene: molecular structure, chromosomal localization, and developmental expression. Dev Biol 127: 45–53

Price JV, Clifford RJ, Schüpbach T (1989) The maternal ventralizing locus *torpedo* is allelic to *faint little ball*, an embryonic lethal, and encodes the Drosophila EGF Receptor homolog. Cell 56: 1085–1092

Priess JR, Schnabel H, Schnabel R (1987) The *glp*-1 locus and cellular interactions in early *C. elegans* embryos. Cell 51: 601–611

Rapp UR, Goldsborough MD, Mark GE, Bonner TI, Groffen J, Reynolds FH Jr, Stephenson JR (1983) Structure and biological activity of v-*raf*, a unique oncogene transduced by a retrovirus. Proc Natl Acad Sci USA 80: 4218–4222

Rijsewijk F, Schuermann M, Wasenaar E, Parren P, Weigel D, Nusse R (1987) The *Drosophila* homolog of the mouse mammary oncogene *int*-1 is identical to the segment polarity gene wingless. Cell 50: 649–657

Rosenthal A, Rhee L, Yadegari R, Paro R, Ulrich A, Goeddel DV (1987) Structure and nucleotide sequence of a *Drosophila melanogaster* protein kinase C gene. EMBO J 6: 433–441

Rothberg JM, Hartley DA, Walther Z, Artavanis-Tsakonas S (1988) *slit*: an EGF-homologous locus of *Drosophila melanogaster* involved in the development of the embryonic central nervous system. Cell 55: 1047–1059.

Rushlow C, Frasch M, Doyle H, Levine M (1987) Maternal regulation of *zerknüllt*: a homoeobox gene controlling differentiation of dorsal tissues in *Drosophila*. Nature 330: 583–586

Sadowski I, Stone JC, Pawson T (1986) A noncatalytic domain conserved among cytoplasmic protein-tyrosine kinases modifies the kinase function and transforming activity of Fujinami sarcoma virus P130^{gag,fps}. Mol Cell Biol 6: 4396–4408

Schechter AL, Stern DF, Vaidyanathan L, Decker SJ, Drebin JA, Greene MI, Weinberg RA (1984) The *neu* oncogene: an *erb*-B-related gene encoding a 185,000-M_r tumour antigen. Nature 312: 513–516

Schejter ED, Shilo B-Z (1985) Characterization of functional domains of p21 *ras* by use of chimeric genes. EMBO J 4: 407–412

Schejter ED, Shilo B-Z (1989) The *Drosophila* EGF receptor homolog (DER) gene is allelic to *faint little ball*, a locus essential for embryonic development. Cell 56: 1093–1104

Schejter ED, Segal D, Glazer L, Shilo B-Z (1986) Alternative 5′ exons and tissue-specific expression of the *Drosophila* EGF receptor homolog transcripts. Cell 46: 1091–1101

Schüpbach T (1987) Germline and soma cooperate during oogenesis to establish the dorsoventral pattern of eggshell and embryo in *Drosophila melanogaster*. Cell 49: 699–707

Schüpbach T, Wieshaus E (1986) Maternal-effect mutations altering the anterior-posteroir pattern of the *Drosophila* embryo. Roux's Arch Dev Biol 195: 302–317

Scott MP, Carroll SB (1987) The segmentation and homeotic gene network in early *Drosophila* development. Cell 51: 689–698

Segal D, Gelbart WM (1985) *Shortvein*, a new component of the *decapentaplegic* gene complex in *Drosophila melanogaster*. Genetics 109: 119–143

Shepard SB, Broverman SA, Muskavitch MAT (1989) A tripartite interaction among neurogenic genes during *Drosophila* imaginal development. Genetics (in press)

Shilo B-Z (1987) Proto-oncogenes in *Drosophila melanogaster*. Trends Genet 3: 69–72

Shimizu K, Nakatsu Y, Sekiguchi M, Hokamura K, Tanaka K, Terada M, Sugimura T (1985) Molecular cloning of an activated human oncogene, homologous to v-*raf*, from primary stomach cancer. Proc Natl Acad Sci USA 82: 5641–5645

Simon MA, Drees B, Kornberg T, Bishop JM (1985) The nucleotide sequence and the tissue-specific expression of *Drosophila* c-*src*. Cell 42: 831–840

Simon MA, Kornberg TB, Bishop JM (1983) Three loci related to the *src* oncogene and tyrosine-specific protein kinase activity in *Drosophila*. Nature 302: 837–839

Spencer FA (1984) The *decapentaplegic* gene complex and adult pattern formation in *Drosophila*. PhD dissertation, Harvard University

Spencer FA, Hoffmann FM, Gelbart WM (1982) *Decapentaplegic*: a gene complex affecting morphogenesis in *Drosophila melanogaster*. Cell 28: 451–461

St Johnston RD, Gelbart WM (1987) Decapentaplegic transcripts are localized along the dorsal-ventral axis of the Drosophila embryo. EMBO J 6: 2785–2791

St Johnston RD, Hoffmann FM, Blackman RK, Segal D, Grimaila R, Padgett RW, Irick H, Gelbart WM (in preparation) The molecular organization of the *decapentaplegic* gene in *Drosophila melanogaster*.

Stahl ML, Ferenz CR, Kelleher KL, Kritz RW, Knopf JL (1988) Sequence similarity of phospholipase C with the non-catalytic region of *src*. Nature 332: 269–272

Steward R (1987) *Dorsal*, an embryonic polarity gene in *Drosophila*, is homologous to the vertebrate proto-oncogene, c-*rel*. Science 238: 692–694

Steward R, McNally FJ, Schedl P (1984) Isolation of the *dorsal* locus of *Drosophila*. Nature 311: 262–265

Steward R, Ambrose L, Schedl P (1985) Expression of the *dorsal* gene. Cold Spring Harbor Symp Quant Biol 50: 223–228

Steward R, Zusman SB, Huang LH, Schedl P (1988) The dorsal protein is distributed in a gradient in early *Drosophila* embryos. Cell 55: 487–495

Technau GM, Campos-Ortega JA (1987) Cell autonomy of expression of neurogenic genes of *Drosophila melanogaster*. Proc Natl Acad Sci USA 84: 4500–4504

Telford BJJ, Butler B, Pirrotta V (1985) Alternative processing and developmental control of the transcripts of the *Drosophila abl* oncogene homologue. EMBO J 4: 2609–2615

Thisse B, Stoetzel C, Dl Messal M, Perrin-Schmitt F (1987) Genes of the *Drosophila* maternal dorsal group control of specific expression of the zygotic gene twist in presumptive mesodermal cells. Genes Dev 1: 709–715

Tomlinson A (1985) The cellular dynamics of pattern formation in the eye of *Drosophila*. J Embryol Exp Morphol 89: 313–331

Tomlinson A, Ready DF (1987) Neuronal differentiation in the *Drosophila ommatidium*. Dev Biol 120: 366–376

Tomlinson A, Boutwell DDL, Hafen E, Rubin GM (1987) Localization of the *sevenless* protein, a putative receptor for positional information, in the eye imaginal disc of *Drosophila*. Cell 51: 143–150

Truman JW, Bate M (1988) Spatial and temporal patterns of neurogenesis in the central nervous system of *Drosophila melanogaster*. Dev Biol 125: 145–157

Vässin H, Bremer KA, Knust E, Campos-Orgeta JA (1987) The neurogenic gene *Delta* of *Drosophila melanogaster* is expressed in neurogenic territories and encodes a putative transmembrane protein with EGF-like repeats. EMBO J 6: 3431–3440

Vincent WS III, Gregory RJ, Wadsworth SC (1989) Embryonic expression of a *Drosophila src gene*: alternate forms of the protein are expressed in segmental stripes and in the nervous system. Genes Dev 3: 334–347

Wadsworth SC, Vincent WS III, Bilodeau-Wentworth D (1985) A *Drosophila* genomic sequence with homology to human epidermal growth factor receptor. Nature 314: 178–180

Weeks DL, Melton DA (1987) A maternal mRNA localized to the vegetal hemisphere in *Xenopus* eggs codes for a growth factor related to TGF-β. Cell 51: 861–867

Wharton KA, Johansen KM, Xu T, Artavanis-Tsakonas S (1985) Nucleotide sequence from the neurogenic locus *Notch* implies a gene product that shares homology with proteins containing EGF-like repeats. Cell 43: 567–581

Wilhelmsen KC, Eggleton K, Temin HM (1984) Nucleic acid sequences of the oncogene v-*rel* in reticuloendotheliosis virus strain T and its cellular homolog, the proto-oncogene c-*rel*. J Virol 52: 172–182

Witte ON (1986) Functions of the *abl* oncogene. Cancer Surv 5: 183–197

Yochem J, Weston K, Greenwald I (1988) The *Caenorhabditis elegans lin-12* gene encodes a transmembrane protein with overall similarity to *Drosophila Notch*. Nature 335: 547–550

Interaction of Oncogenes with Differentiation Programs

David Boettiger

1 Introduction

One recurrent theme in the exploration of the function and mechanism of action of oncogenes and their proto-oncogene forebears is to view these genes in the

Department of Microbiology, University of Pennsylvania Philadelphia, PA 19104-6076, USA

Current Topics in Microbiology and Immunology, Vol. 147

context of normal development. There is both direct evidence that oncogenes can alter the expression of marker proteins in a variety of differentiated cell types and evidence for the developmental regulation of expression for many proto-oncogenes. Taken together this evidence leads to the speculation that some of these proto-oncogenes function as regulators of normal cell differentiation. Unfortunately, our current knowledge of the mechanisms which control cell differentiation are rather primitive. In this context, the focus is one the control of cell fates during development and the regulation of large groups of gene products which serve to distinguish different cell phenotypes. While this process is likely to share some properties with the systems which regulate individual genes, the question is more one of the coordination of gene functions, what controls or coordinates the regulators, what allows them to respond to developmental signals, and what is the nature of the signals to which they respond. It is likely that these questions will have a multiplicity of answers which will depend on the cell type under consideration. It is my belief that the studies on the functions of oncogenes will provide critical insights on the mechanism underlying the mysteries of cell differentiation. My essential purpose in this review is to redefine some basic developmental concepts in a form which provides a context for discussion of the developmental roles for the proto-oncogenes and for the intervention in the developmental process by oncogenes.

2 Developmental Programs

2.1 Social Functions of Cells

The function of development is the creation of a multicellular organism in which individual cells performs specialized functions which are critical to the overall survival of the organism. Since this specialization involves cells of the same genotype that originate from a single cell, the specialization cannot be totally programmed internally but must involve information received from the cell's microenvironment. The processes of cell differentiation and cell proliferation in multicellular organisms are essentially social functions. In the normal individual, these processes only take place in the context of other cells and thus it is critical for signals to be passed between cells in order to maintain social coordination. Oncogenesis or neoplasia is essentially a breakdown of this normal social coordination. Hence it is likely that genes whose protein products function in this process could, in a mutated form or due to mis-regulation, serve as oncogenes. This supposition is clearly borne out in the identification of oncogenes which were derived from genes for both peptide hormones (growth factors) and their cellular receptors (Bishop 1985). In normal circumstances, the growth factors are produced by one set of cells and control the proliferation (and other responses) of another cell population, thus serving to communicate normal social regulation. These peptide hormone and hormone receptor oncogenes are currently the best understood groups of oncogenes in terms of their physiological function and their relationship to normal cell physiology. It seems reasonable to propose that many of the remaining oncogenes also function in social behavior of cells. This hypothesis appears to be supported by the absence of homologues for most oncogenes in unicellular organisms such as yeast (with the exception of the *ras* homologs) (Fukui et al. 1986; J. B. Levy and J. M. Bishop, personal communication) despite their wide distribution in multicellular organisms (Weiss et al. 1982).

The social functions of cells require information transfer from the environment to the cell regulatory machinery so that an appropriate response is achieved. This is generally accomplished via cell surface molecules which act as receptors. Although the best understood members of this group are the peptide hormone receptors, new classes of receptors are now being defined which are involved in cell-cell interactions and in cell-extracellular matrix interactions. These include the various cell adhesion molecules (CAMs; Edelman 1986) and the integrin family of extracellular matrix receptors (Buck and Horwitz 1987; Hynes 1987). The described systems probably represent the tip of the iceberg. Many of these interactions have been described in terms of adhesion, but recent experiments have demonstrated that they also function in signal transduction (Menko and Boettiger 1987). Hence one can speculate that some of the known oncogenes sould be involved in the regulation of these "adhesion" events and transmission of the signals generated. This is quite reasonable in the light of the fact that it is the adhesion of cells to substrate that is a primary determinant of cell morphology in culture (and probably even more so in vivo), and transformation of these cells resulting from the expression of a viral oncogene is accompanied by altered cell morphology.

2.2 Definitions and Developmental Models

In classical genetics cells are discussed in terms of the cell genotype encompassing the total coding of the cell DNA and cell phenotype encompassing the total cell products exclusive of the DNA. For discussion of cell differentiation, it is useful to interpose an intermediate level which I call the cell program. This program is what distinguishes the potentialities of two different cell types within the same organism; it allows different cells to respond differently to the same stimulus. That such differences exist is demonstrated by the fact that cells put into tissue culture under similar conditions may express different phenotypes, and that different cell types respond differently to stimulation by the same hormone (Massague 1987). The cell program is distinct from the cell phenotype since it is possible for two cells with the same differentiation program to express different phenotypes if they have been exposed to different external stimuli (microenvironments), or the cell program may be expressed in one cell and not expressed by another cell. It is a developmental cognate of the genetic concept of genotype. The program consists of genes which are expressed, genes which can be induced by various means, and genes which are effectively inaccessible. Presumably, this program is "written" in the chromatin-DNA structure of the cell and generally propagated to the daughter cells at cell division.

During the course of normal development, there must be changes in the cell programs of individual cells. The result of these programming changes is to restrict the cell's potentialities, i.e., the functional cell types of daughter cells which can be produced. A simplified model of cell differentiation and cell programming is given in

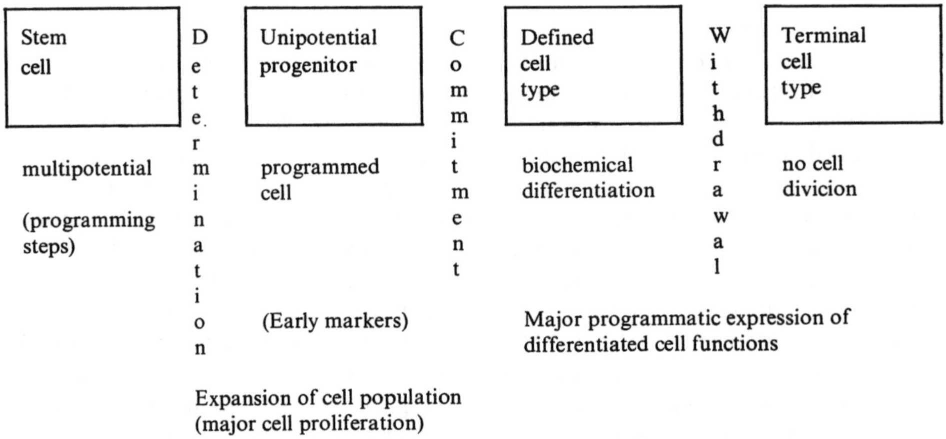

Fig. 1. Developmental compartments

Fig. 1. It begins with stem cells or multipotential cells which can produce descendants of two or more distinct phenotypes and which include cells from the zygote to the stem cells present in bone marrow, gut, and skin in the adult animal. Eventually this compartment may be meaningfully subdivided as we acquire more information about the cells. The next compartment contains cells which are determined or have a developmental program restricted to a single developmental lineage and will produce only a single progeny cell type. The third compartment is the initiation of expression of the program and synthesis of cell products characteristic of the terminal phenotype of the cell lineage. During one or more of these stages there is considerable cell proliferation to give a sufficient population of terminally differentiated cells. Finally, in some lineages there is a further stage of differentiation which requires permanent withdrawal from the cell cycle. The virtue of this simple model is that it provides operational definitions for each of the transitions from one compartment to another and presents a set of compartments based on the state of the cell program.

2.3 Oncogenes and Developmental Programs

The effects of oncogenes on the expression of differentiated cell products could operate directly on the expression of the individual genes which encode these products, or their effects could be mediated by the particular cell program. Two lines of evidence support the hypothesis that, at least in the case of the *src* oncogene, the effect is on the cell program rather than directly on the individual genes. First, expression of the *src* oncogene in fibroblasts results in a suppression of synthesis of fibronectin which is mediated at the RNA level, whereas in chondroblasts expression of the *src* oncogene increase the level of *src* mRNA (Allebach et al. 1985). In these cases the same stimulus exerted by the expression of the *src* oncogene has the opposite effects on fibronectin messenger (m)RNA synthesis, indicating that the cells are programmed differently. Second, *src* oncogene expression suppresses whole groups of genes which encode most of the major differentiated cell products in both chondroblasts (PACIFICI et al. 1977; SHANLEY et al. 1983; ADAMS et al. 1982) and myoblasts (FISZMAN and FUCHS 1975; ALEMA and TATO 1987; HOLTZER et al. 1975; ANTHONY et al. 1984). This coordinated effect suggests that this suppression may be mediated by the cell program which has set up these genes for efficient expression in these cell types. The pattern of affecting a whole collection of genes which encompasses much of the specific cell program for particular differentiated cell types is repeated in the case of other oncogenes, although many of these cases have not been studies in as much detail. These include suppression of erythroid cell transformation by *erbB* (SAMARUT and GAZZOLO 1982; RAYNAUD et al. 1987; BEUG et al. 1982a), suppression of macrophage differentiation of *myb* (GAZZOLO et al. 1979; BEUG et al. 1979; SHEN-ONG et al. 1987; DURBAN and BOETTIGER 1981a), and suppression of myogenesis by *erb*B (FALCONE et al. 1985) or *fps* (FALCONE et al. 1985).

In most experiments, oncogenes have been introduced into cells which already express their biochemical differentiation. As discussed above, in many of these cases it appears that the effect of the oncogene on expression of individual proteins depends on the cell type and thus is governed by the cell program. When temperature-sensitive mutants of oncogenes have been used a shift to nonpermissive temperature generally results in the reexpression of the normal differentiated cell phenotype (HOLTZER et al. 1975; ADAMS et al. 1982; PACIFICI et al. 1977; MENKO and BOETTIGER 1988; BOETTIGER et al. 1977). Thus the normal cell program does not appear to have been affected by the oncogene. By infecting cells at earlier stages of differentiation, particularly at the stem cell level, it may be possible to produce a situation in which the programming steps (see Fig. 1) take place in a cell expressing the oncogene. In this case will the oncogene affect the cell program? There have been a number of attempts to get oncogene expression in stem cells and there is no evidence that the expression of an oncogene has any effect on the stem cell itself or on its ability to produce differentiated cell progeny. In most cases differentiated cell progeny which expressed the oncogene did transform. This pattern has been born out for several systems including transformation of chondrogenic cells (BOETTIGER et al. 1983) and embryonal carcinoma cells (BOULTER and WAGNER 1988) by v-*src*, transformation of macrophage lineage cells by v-*myb* (BOETTIGER and DURBAN, 1980; OLSEN and BOETTIGER, in preparation), transformation of erythroid cells by *erb-B* (SAMARUT and GAZZOLO 1982), and transformation of embryonal carcinoma stem cells by *ras* (BELL et al. 1986). The use of temperature-sensitive mutants in some systems (BOETTIGER et al. 1983) allowed determining whether the cells did differentiate by shifting the transformed cells to the non-permissive temperature.

The preliminary conclusion from these experiments is that the several oncogenes which were tested do not affect the process of cell programming; however, it is still possible that oncogene expression only occurred following the programming step as we do not know exactly when programming occurs. These experiments bear out the original hypothesis that oncogenes can be introduced into progenitor cells but that transformation only occurs in the differentiated progeny (BOETTIGER and DURBAN 1980).

3 Oncogenes and Proto-oncogenes: Developmental Specificity

The effects of oncogene expression on the cell phenotype are determined in part by the particular cell type in which the oncogene is expressed. Expression of an oncogene in a particular cell type may have no effect, or may produce a particular transformed cell phenotype (Durban and Boettiger 1981a). Analysis of the expression of proto-oncogenes has determined that their expression varies with the differentiated state of the cell (Adamson 1987; Muller and Verma 1984; Bishop 1983). Developmental specificity is this specificity of oncogenes for expression and action in particular cell types. The ability of an oncogene to cause an effect or to transform a cell is related to the presence of particular target molecules with which it must interact. These may form part of a cascade of biochemical reactions which lead to phenotypic changes or an intracellular signalling pathway which can be disrupted. Developmental specificity has been assayed using two general approaches: (1) introduction of an exogenous oncogene either using a retrovirus or some other form of gene transfer, and (2) assay of the expression of proto-oncogenes in different cell types. While each of these approaches has different limitations, both address the question of which cell types express the appropriate pathway(s) for interaction with particular oncogenes. The introduction of exogenous oncogenes is more direct and is expected to reveal a wider range of cell types in which a particular oncogene may function. The expression of a proto-oncogene at a particular point in development or in particular cell types suggests that the appropriate pathways for its normal function are likely to be present.

3.1 Cell Type Specificity for Oncogene-Induced Transformation

The first indications of developmental specificity for oncogenes came from studies on the pathogenesis of naturally occurring retroviruses which encode oncogenes. The results of such studies are available in a number of reviews on the pathogenesis of retroviruses (Beard 1963; Weiss et al. 1982; Gross 1970; Graf and Beug 1978). This approach of using the pathogenesis of retroviruses provides evidence for cell types which can respond to particular oncogenes as long as one focuses on the common tumor types which are produced with short latent periods. But the results are limited by the restrictions imposed by cells which are susceptible to infection and the anatomical and physiological restraints in delivering a sufficient virus dose to particular tissues. These studies have been most informative for hematopoietic cell types where the anatomical constraints are minimal. The range of cell types shich can respond to individual oncogenes has been significantly expanded through the use of tissue cultures systems and more recently with transgenic animals carrying particular oncogenes (the transgenics are discussed in a later section). In defining the range of cell types which can be affected by a particular oncogene, it is as important to define cell types which cannot respond as it is to identify cell types which can respond. The identification of the nonresponder cell types is a more difficult problem and it is important to eliminate trivial reasons for failure to respond.

Avian myeloblastosis virus which carries the *myb* oncogene provides the best example of a cell type host range which is developmentally restricted both in terms of cell lineage and stage (WEISS et al. 1982; GRAF and BEUG 1978). While there was some confusion in the initial identification of the tumor cell type which led to a misnaming of the virus, it is now clear that the in vivo and in vitro transformed cells are members of the monocyte lineage (DURBAN and BOETTIGER 1981a; NESS et al. 1987). Experiments with other cell types, including other hematopoietic cells (BEUG et al. 1979; GAZZOLO et al. 1979; BOETTIGER and OLSEN 1989) and fibroblasts (KLEMPNAUER et al 1984), indicate that the effect of the virus is restricted to the monocyte lineage. Through use of the hematopoietic progenitors present in the avian yolk sac, it has been possible to demonstrate that the effect of this viral oncogene is limited to developmental stages beginning at the promonoblast, i.e., after the yolk sac monocyte progenitor (BOETTIGER and DURBAN 1984; BRELVI and STUDZINSKI 1987). Thus, there is a specificity for both the monocyte lineage and for the more mature stages of differentiation within the lineage. However, other experiments using the *myb* oncogene indicate that other hematopoietic lineages may also be affected.

A series of chicken B-cell tumors have been identified in which c-*myb* is activated by promoter insertion, which implies that the B cells also respond to *myb* (KANTER et al. 1988). *myb* also appears to affect cells in the erythroid lineage. Another retrovirus, E26, which contains a *myb* oncogene transforms both monocyte and erythroid cells (RADKE et al. 1982; MOSCOVICI et al. 1983), although it is possible that the erythroid specificity is conferred by a second oncogene, *ets*, in this virus (BISTER et al. 1982; WATSON et al. 1985; NUNN et al. 1984). Introduction of c-*myb* into Friend erythroleukemia cells blocks their ability to differentiate in response to differentiation inducers (CLARKE et al. 1988a). Hence the developmental specificity of *myb* includes both the erythroid and B-cell lineages.

In contrast to the restricted range of cell types which respond to *myb*, *src* can effect a broad range of cell types. It is able to transform a wide variety of cells of both mesodermal and epidermal origin (PACIFICI et al. 1977; FISZMAN and FUCHS 1975; BOETTIGER et al. 1977; MENKO and BOETTIGER 1988), and even erythroid cells (KAHN et al. 1986). But even *src* is not able to transform all cell types. Macrophages in which the viral *src* oncogene is expressed at high levels and is properly localized to the cell membrane do not display either altered cell morphology or altered cell growth (DURBAN and BOETTIGER 1981a; LIPSICH et al. 1984). Like *myb*, the effects of *src* appear to be primarily on the expression of differentiation markers which are expressed at a late stage in the lineages affected by *src* (discussed above). Thus *src* does demonstrate a developmental specificity.

The dependence of developmental specificity on the presence of specific intracellular signalling pathways is illustrated by the *fms* oncogene. It was identified as the transforming gene of the McDonough strain of feline sarcoma virus which also transforms fibroblasts in vitro (MCDONOUGH et al. 1971; ANDERSON et al 1982). The cellular homologue for *fms* is the receptor for colony-stimulating factor 1 (CSF-1), a factor involved in the differentiation and proliferation of macrophages (TUSHINSKI et al. 1982). As would be expected, if this oncogene delivers a constitutive CSF-1 receptor signal, the viral *fms* oncogene is able to transform macrophages (WHEELER et al 1986). These results suggest that a common intracellular signalling pathway which can be initiated by *fms* is present in both macrophages and fibroblasts even though they normaly respond to different exogenous factors. This hypothesis was further confirmed by the introduction of the human c-*fms* gene (CSF-1 receptor) into NIH 3T3 cells. The proliferation of these cells can now be controlled by the addition of human CSF-1 (ROUSSEL et al 1987). Thus the introduction of an exogenous oncogene may be used as a means for determining the presence of certain intracellular signalling pathways.

For oncogenes which appear to have their major effect in the suppression of cell differentiation, the developmental specificity may be interpreted in terms of developmental regulation. However, it is possible that some oncogenes function primarily in the regulation of the cell cycle and act in most cell types on the cell cycle regardless of the differentiated phenotype of the cell. In such a case, the oncogene would be expected to affect the differentiation of the cell when it is closely tied to the cell cycle. Both erythrocytes and muscle cells must withdraw from the cell cycle as a prerequisite for terminal differentiation, and hence these differentiation processes might be blocked by a cell cycle oncogene. The *myc* oncogene may be a member of such an oncogene class. It appears to affect a wide variety of different cell types. although its transforming function or tumorigenicity may require the participation of a second oncogene (LAND et al. 1986; LEDER et al. 1986). Furthermore, its expression has been reported to block both erythroid (DMITROVSKY et al. 1986b) and myogenic differentiation (FALCONE et al. 1985).

3.2 Expression of Proto-oncogenes

Proto-oncogenes are the cellular homologues which are the progenitors to the transforming oncogenes. They differ from their oncogenic counterparts usually as a result of mutations or rearrangements in controlling sequences which may affect either the presence of cellular mRNA or the regulation of function of the protein product. The expression of these genes is generally more restricted than the susceptibility to transformation discussed in the previous section. There is a rather extensive literature on the expression of proto-oncogenes. Much of the data is based on whole organs or groups of organs, or whole embryos at different developmental stages. However, the actual level of expression in different cell types within these tissues can vary drastically. The expression of c-*myb* in yolk sac hematopoietic cells has been shown by cell fractionation to be limited to about 5% of the cell population (DUPREY and BOETTIGER 1985). In situ hybridization for c-*myc* (PFEIFER OHLSSON et al. 1985) and immunostaining for c-*src* in neural tissue (SORGE et al. 1984) have also demonstrated cell type differences in proto-oncogene expression within a tissue. Thus identification of the cell types rather than tissue types in which proto-oncogenes are expressed is critical to understanding their function in the developmental context.

3.2.1 Proto-oncogene Families

Most of the proto-oncogenes are members of related gene families (see also Sect. 5). There are many other gene families, including those for isoenzymes for many enzymes, in which different family members show different developmental expression. Similar differential regulation has been defined in a number of the proto-oncogene families. The trends are clear on the basis of comparison of organs and developmental stages for *src* and *yes* (GESSLER and BARNEKOW 1984): K-*ras*, H-*ras*, and N-*ras* (LEON et al. 1987); and for c-*myc*, N-*myc*, and L-*myc* (ZIMMERMAN et al. 1986). As for other developmental markers, it is not clear whether this differential expression of the family members reflects functional differences in the protein product which affects its interaction with other cell components, or whether it reflects the organization of the developmental program at the gene level.

3.2.2 Stage-Specific Expression

The stage within a developmental lineage when a proto-oncogene is expressed is an important issue. The timing of expression will influence the effect which the proto-oncogene may have on the developmental program. Since expression of proto-oncogenes is often found in tumor cells (see below) and introduction of oncogenes can cause cell transformation, one might expect that cells at early stages in the various cell lineages which express limited cell differentiation and high proliferative potential might also express these proto-oncogenes. This pattern does appear to hold for some oncogenes. c-*myb* is expressed at early stages in the monocyte (DUPREY and BOETTIGER 1985), erythrocyte (EMILIA et al. 1986; KIRSCH et al. 1986). B cell (KAN et al. 1985), and T cell (SHEINESS and GARDINIER 1984a) lineages. Its expression drops by up to two orders of magnitude as the cells pass a certain point of maturation within the monocyte lineage (DUPREY and BOETTIGER 1985). The *int-1* oncogene shows a very specific expression in early neural tube of the mouse embryo (WILKINSON et al. 1987; GIOTTA et al. 1980). In contrast, a number of proto-oncogenes increase in expression as differentiation progresses. In macrophage differentiation, while c-*myb* decreases, there is an increase in c-*fms* (the receptor for its major growth factor) (SARIBAN et al. 1985; WILLMAN et al. 1987), c-*fgr* (WILLMAN et al. 1987), c-*fps* (MACDONALD et al. 1985; FERRARI et al. 1985a; FELDMAN et al. 1985), and c-*sis* (MARTINET et al. 1986). Other proto-oncogenes exhibit a dual pattern of expression, first in the early embryo and later accumulated in specific differentiated tissues of the adult. The most direct example is c-*src*, which exhibits an early peak of expression (GESSLER and BARNEKOW 1984; SCHARTL and BARNEKOW 1984) that includes the neural ectoderm at the stages of gastrulation and early neural tube formation (MANESS et al. 1986a). This is followed by a decrease of c-*src* levels in the neural tissues and later reappearance and accumulation in the fully differentiated neurons (MANESS 1986b; SORGE et al. 1985). However, the c-*src* expressed in these late neural tissue encodes a slightly different protein due to an alternative mRNA splice in neural tissue (LEVY et al. 1987). Other oncogenes which are usually associated with proliferating cells also accumulate in late stages of differentiation in postmitotic muscle and brain. c-*myc* is expressed at high levels during the early phases of development and is present as maternal mRNA in the ovum (KING et al. 1986), but it accumulates in the adult in muscle (ENDO and NADAL GINARD 1986). *ras* has also been reported to accumulate in adult muscle and brain (LEON et al. 1987).

3.2.3 Proliferation vs Differentiation

Based on their association with tumors, it has been proposed that proto-oncogenes would be associated with processes of cell proliferation. Hence, one would expect a distribution of expression in which higher levels of expression would be associated with proliferating cells. While this might be true in the case of tumor tissue (see below), no proto-oncogenes have been found whose distribution is limited to proliferating cells. In situ hybridization experiments performed on fetal tissue have not been able to demonstrate a preferential association of either *ras* (Mellersh et al. 1986) or *myc* (Pfeiffer Ohlsson et al. 1985) with proliferating cells. These genes were examined since their association with cell proliferation and tumor cells is most often cited. Hence it appears that proto-oncogene expression is determined largely by the differentiation program of the cell and less by its proliferation status. This conclusion is not too surprising since cell proliferation itself is part of the developmental program and is controlled by many factors during development.

3.2.4 Proto-oncogene Induction

The proto-oncogenes c-*fos* and c-*myc* are rapidly induced in response to a variety of membrane signalling events. The original observation that c-*myc* and c-*fos* were among the rapidly induced genes following exposure of BALB 3T3 cells to platelet-derived growth factor (PDGF) has been followed up using a variety of inducers (Kruijer et al. 1984). In some cases the induction may not include both c-*myc* and c-*fos*, but coinduction appears to be more common. Induction occurs in response to the growth factors: interleukins IL-1a (Kovacs et al. 1986), IL-2 (Reed et al. 1985, 1987; Shipp and Reinherz 1987), and IL-3 (Conscience et al. 1986; Umemura et al. 1988); epidermal growth factor (EGF) (Bravo et al. 1985; Dolberg and Bissell 1984), PDGF (Wome et al. 1987; Kruijer et al. 1984), B-cell growth factor (BCGF) (Lacy et al. 1986), insulin-like growth factor 1 (IGF-1) (Ong et al. 1987), nerve growth factor (NGF) (Milbrandt 1986; Kruijer et al. 1985; Curran and Morgan 1985), and granulocyte-macrophage colony stimulating factor (GM-CSF) (Jaffe et al. 1988). Induction occurs in response to hormones which regulate differentiation and physiological responses: thyroid-stimulating hormone (TSH) (Dere et al. 1985), thyrotropic hormone (Colletta et al. 1986), and erythropoietin (Umemura et al. 1988). Also, a number of chemicals which interact with membrane stimulatory events can serve as inducers. These include isoproterenol which is a β-adrenergic agonist (Barka et al. 1986), calcium ionophores (Chapekar et al. 1987; Lacy et al. 1986), retinoic acid (Chapekar et al. 1987), and the tumor promoter TPA (Nilsson et al. 1986; Skouv et al. 1986). Stimulators of the immune response including antigen (Chapekar et al. 1987; Lacy et al. 1986) and the mitogens lipopolysaccharide (LPS) (Introna et al. 1986) and concanavalin A (Reed et al. 1987) induce c-*myc* and c-*fos* expression in B cells, T cells, and macrophages. While this oversimplifies the interactions, this wide collection of stimuli which c-*myc* and/or c-*fos* suggest that the products of these genes are involved in the cellular responses to membrane signalling events. What is the role of these responses which are shared by so many different stimuli?

In addition to the rapid induction of c-*myc* and c-*fos*, there is increasing evidence that this early transient expression of c-*myc* and c-*fos* is (or can be) followed by a later peak of expression which can involve c-*myb*, c-*myc*, or c-*fos* (Reed et al. 1987; Giroldi et al. 1988; Jaffe et al. 1988).

3.2.5 Patterns and Function

One objective of following the cell type-specific expression of the proto-oncogenes is to obtain clues about its mechanism of action. While this approach has not been as successful as the sequencing approach, it has provided an important adjunct to those studies. The evidence that c-*fms* expression was limited to late myeloid cells (SARIBAN et al. 1985) led to its identification with the CSF-1 receptor (SHERR et al. 1985). Perhaps because our understanding of their mechanism of action is more advanced, the expression of both growth factors and growth factor receptor oncogenes is more predictable. Other patterns of proto-oncogene expression are more difficult to interpret in our current framework. For example, what is the common function for c-*src*, based on its elevated expression in early embryonic cells (SCHARTL and BARNEKOW 1984), platelets (GOLDEN et al. 1986), and neurons (BRUGGE et al. 1985)?

Some of the initial questions raised in the studies on proto-oncogene expression patterns have been answered. First, there are developmental patterns of expression. Second, there are proto-oncogenes which are expressed early and hence may affect the cell programming or commiment steps. Third, no proto-oncogenes have been found whose expression is correlated with cell proliferation.

3.3 Expression in Tumors

There is an expanding literature on the elevated expression of proto-oncogenes in tumor cells. This elevated expression could reflect the normal level of expression in the founder cell for the tumor and hence be unrelated to the tumorigenic process, or it could be increased as part of the tumorigenic process. In the former case, the identification of tumors with elevated expression provides additional information on the normal pattern of proto-oncogene expression. Unfortunately, it is difficult to determine the exact developmental stage of the founding cell for the tumor and thus there is no appropriate normal control. However, if one examines the expression patterns for relatively restricted proto-oncogenes such as c-*myb*, the bulk of the tumors expressing elevated levels are derived from hematopoietic cells (SLAMON et al. 1986; FERRARI et al. 1985b; YOKOTA et al. 1986; ROY BURMAN et al. 1983) as would be expected from the normal pattern of expression of c-*myb* (see above). But c-*myb* may also be expressed at elevated levels in tumors derived from cells which probably do not have elevated levels prior to transformation, as indicated in the occasional reports of elevated levels in carcinomas (GRIFFIN and BAYLIN 1985). This suggests that c-*myb* was activated as part of the tumorigenic process but it is not clear whether the activation is a cause or the result of the tumorigenic phenotype. Is it an accident that elevated levels of c-*src* are rarely found in tumors, or is this a reflection of the restriction of its normally elevated expression to postmitotic cells in the adult (see above)? Thus, while data from expression of proto-oncogenes in tumor cells can be important in identifying the normal cells which express the proto-oncogene, there are some exceptions.

The other possibility is that the elevated expression is related to the tumorigenic process. While this is likely to be true in some cases, as in the unusual cases of c-*myb* elevation, this situation will reveal less about the developmental interactions of the particular oncogene until it is clear that the elevated level is important in the tumorigenic process and that it is not just a result of lapses in normal developmental control which accompany tumorigenesis. The argument for a causative role is easier to make for oncogenes which are regularly found active in a particular tumor type. The particular evolvement of specific proto-oncogenes in particular tumor types should emerge as the patterns of expression are better defined.

4 Approaches, Limitations, and Controls

The function of oncogenes in cell proliferation has become the major focus for studies on oncogene function at the cellular level. The reason for this is not that increased cell proliferation is the most or sole important characteristic of tumor cells, but that tissue culture systems for study of cell proliferation are better developed and the end point of cell proliferation is more easily defined experimentally. In neoplasia, it is not only the proliferative potential of the cell but also its differentiated state which is disturbed. In fact, the pathological classification of cancers is based on morphological (i.e., differentiation) criteria. The objective of this section is to discuss experimental differentiation systems and their advantages and limitations for analysis of oncogene function and oncogene perturbation of development.

4.1 Primary Culture Systems

The problem of analyzing the effect of oncogene expression on cell differentiation is complex. Cell differentiation is affected and controlled by hormone systems, cell-cell interactions, cell-matrix interactions, and cell-intrinsic patterns of sequential gene expression. While we understand some aspects of each of these factors, much information is missing particularly as it relates to the coordination of these processes in embryogenesis. Tissue culture offers a means to isolate and manipulate parts of this total system. The types of culture systems can be defined on the basis of the developmental model in Fig. 1.

4.1.1 Defined Cell Types

Most of the experiments which have examined the effects of oncogenes on cell differentiation have used cultures of defined cell types. These cultures are produced by a combination of dissection, developmental staging, selective dissociation procedures, and selective culture conditions. The critical properties of these cultures for use in studies using oncogenes include: (a) cell-type purity, which is critical for the control of cell-cell and cell-extracellular matrix interactions and for molecular analysis of the level of action of the particular oncogene; (b) ability to divide, which is essential for proper expression of introduced genes and desirable for some population expansion; and (c) the ability to retain expression of the normal differentiated cell phenotype under our artificial culture conditions. Experiments with differentiated cell types are only as good as the stringency with which these parameters are maintained. The imposition of the second criterium and the introduction of foreign genes provide further stress on the system and hence it is often necessary to improve culture procedures commonly used for less demanding studies. This problem has been encountered in the experiments on chondroblasts (PACIFICI et al. 1977) and in limb bud cultures in which suppression of differentiation can be due to culture conditions rather than the oncogene (BOETTIGER et al. 1983; GROSS and RIFKIN 1979). Given the problems associated with the maintenance of normal developmental expression under varying culture conditions it is particularly important to pay attention to controls. Whenever possible the use of a conditional oncogene expression system is desirable (e.g., use of temperature-sensitive oncogene mutants). In cultured thyroid cells the loss of developmental gene expression following transformation by an oncogene did not revert on inactivation of the oncogene by a temperature shift (FUSCO et al. 1985). Thus it is likely that the loss was due to the culture conditions rather than the oncogene expression. This effect of loss of expression is enhanced by cloning of the cells following introduction of the oncogene. In cells transformed by a temperature-sensitive mutant of *src*, cloning in agar leads to loss the ability to differentiate into myotubes following a temperature shift in many of the clones (S. ALEMA, F. TATO, and D. BOETTIGER, unpublished results). Another technical problem is that these differentiated cell populations are often more difficult to infect with retroviruses (PACIFICI et al. 1977; BOETTIGER et al. 1977). Part of this problem is explained by the presence of extracellular matrix which may protect the cell receptors from the virus.

The most common primary culture systems are based on the chicken embryo because of ease of obtaining tissues and the existence of a large base of developmental studies using these embryos. Chondroblasts, dermal fibroblasts, and tendon fibroblasts can be obtained in greater than 98% pure populations in culture (PACIFICI et al. 1977; MENKO et al. 1983; SCHWARTZ et al. 1978). We have used the chondroblast system extensively for studies on the molecular mechanisms of differentiation suppression by the *src* oncogene (ADAMS et al. 1982; ALLEBACH et al. 1985). The avian myogenic system is the most widely used of these "pure culture" systems (FISZMAN and FUCHS 1975; ALEMA and TATO 1987; HOLTZER et al. 1975; ANTHONY et al. 1984). In this system, the starting population consists of proliferating myoblasts which are contaminated with nonmyogenic cells and both the myogenic and nonmyogenic cells have "fibroblastic" morphologies in culture and cannot be easily distinguished. Thus the level of purity of an individual culture has been difficult to determine. The monoclonal antibody L4 distinguishes the myogenic cells in these cultures now making it possible to evaluate purity (GEORGE-WEINSTEIN et al. 1988; MENKO and BOETTIGER 1987). This is important since purity can vary widely using published preparation procedures. Under optimal conditions, purity of about 90% can be achieved. In myogenic differentiation, replicating cells express few muscle-specific markers and withdrawal from the cell cycle is accompanied by the synthesis of the major muscle-specific proteins and fusion into myotubes (HOLTZER et al. 1983; ALEMA and TATO 1987). This sharp distinction of compartments and the link of cell division and differentiation has made this a particular sensitive test for oncogene effects (discussed in detail below, and see recent review: ALEMA and TATO 1987).

The culture of epithelial cells has lagged behind that of mesodermally derived cells. This is probably due to the greater complexity of the cell-cell and cell-matrix interactions which are required for the maintenance of their normal phenotype. However, significant progress is being made in several systems. The mouse primary keratinocyte culture system cells differentiate in response to Ca^{2+} with production of keratins and elaboration of a cornified epithelium. The calcium-dependent differentiation is suppressed by the oncogenes *abl*, *src*, *fes*, *ras*, and *mos* (YUSPA et al. 1985; WEISSMAN and AARONSON 1985; RHIM et al. 1985). Pigmented retina epithelial cells can retain their ability to accumulate melanin and produce melanosomes in culture. Expression of the *src* oncogene in these cells results in the loss of melanin and the destruction of the melanosomes (BOETTIGER et al. 1977). Cultures of chick lens epithelial cells will differentiate in vitro to form lentoid bodies with a three-dimensional structure resembling that of a normal lens. The process involves formation of extensive gap junctions, production of a basal lamina, and synthesis of high levels of δ-crystallin (MENKO et al. 1984). The formation of these lens structures is prevented by expression of the *src* oncogene (MENKO and BOETTIGER 1988). Neural retina cells expressing the *src* oncogene loose some structural characteristics of neural cells but retain the ability to synthesize γ-aminobutyric acid (GABA) and several other neural proteins (HALL et al. 1987; CRISANTI COMBES et al. 1982; NOTTER et al. 1987). The neural retina cultures contain a mixture of cell types and only limited attention has been paid to expression of developmental markers; rather, attention has focused on the extended proliferation induced by oncogene expression. Cells derived from the cerebellum of a rat embryo have been infected by a virus carrying the *src* oncogene and a clone of "transformed" neural cells obtained (GIOTTA et al. 1980). Some properties of differentiated neurons appear to be lost and other differentiation markers are still expressed (GIOTTA and COHN 1981). In these studies it is difficult to sort out the effects of the oncogene from those of the culture conditions.

In summary, these culture systems consist largely of cells which already express most of the products associated with their terminal phenotypes at the time when the oncogene has been introduced. In general, there is a suppression of the expression of the differentiated cell products; however, not all differentiation-specific products are suppressed. In addition to the loss of synthesis of these products, some cells actually destroy structures and cell organelles in response to oncogene expression.

4.1.2 Stem Cells and Progenitor Cells

In addition to effects on the differentiated cells in the defined cell type and terminal cell compartments, it is possible that oncogenes will have effects on earlier stages of differentiation. The future development of model culture systems could permit analysis of the possible action of oncogenes on the processes involved in cell programming. Analysis of the effects of oncogenes on these early stages of differentiation is more difficult because there are limited numbers of these cells in the body (for example, even in bone marrow the hematopoietic stem cell represents only 1 in 10^4 cells) and there are no well-defined, unique markers to distinguish individual cell types. Instead these progenitor cells are distinguished by the progeny they produce. This could be reflected either in the particular progeny cell types as a reflection of their potentialities or in the size of the progeny cell population as a measure of their self-renewal. (Self-renewal is cell proliferation without change in cell potentialities, i.e., the daughters are like the parent.)

Hematopoiesis currently offers the best system to approach these problems. From the practical point of view, a series of clonal assays using semisolid media has been developed for different hematopoietic progenitor cell types (METCALF 1977; DEXTER 1984). These assays allow both quantitation of the founder cell population and the ability to classify individual colonies on the basis of the differentiated cell types which it contains. The specificity of these assays is determined primarily by the particular hematopoietic growth factors which are added to the culture (DEXTER 1984). With the cloning of the genes which encode most of these factors in the murine and human systems and the availability of pure factors as a result of this cloning (METCALF 1986), the specificity of these assays is more precisely defined. Using these assay systems means that the progenitor cells are defined on the basis of the factors to which they are able to respond and that an individual cell may respond to more than one factor. Certainly, cloned hematopoietic cell lines which respond to more than one factor have been described (IKEBUCHI et al. 1988; HARA et al. 1988). Thus, in spite of our best efforts to analyze normal hematopoiesis lineage structure with these tools (OGAWA et al. 1983; JOHNSON 1984), the branching structures proposed within the progenitor cell compartments are ambiguous. Such ambiguities are more often resolved by theoretical bias, such as the limitation of a cell to binary decisions (JOHNSON 1984) than by experimental disproving of alternate hypotheses. In the avian system, macrophage and erythroid colony assays have been utilized to analyze the series of acute avian leukemia viruses carrying *myb*, *myc*, *erbB*, *erbA* oncogenes (BOETTICHER and DURBAN 1984; GAZZOLO et al. 1979; RADKE et al. 1982; GRAF et al. 1981; SAMARUT and GAZZOLO 1982). In the murine system these assays have focused on B-cell differentiation for analysis of the effects of the *abl* oncogene (WITTE 1986; ROSENBERG and BALTIMORE 1976; WONG et al. 1988), and on the erythroid system in the analysis of effects of *ras* (HANKINS and SCOLNICK 1981) and *abl* (WANECK and ROSENBERG 1981).

In addition to the colony assay approach, cultures have been produced from primary hematopoietic tissue such as bone marrow and yolk sac. These have been used in short-term cultures in which there is essentially a single wave of differentiation (BOETTIGER and DURBAN 1980; IHLE 1986) or in long-term bone marrow cultures which have been developed for their ability to maintain the viability, self-renewal, and differentiation of the hematopoietic stem cells in vitro (GREENBERGER 1979a; DEXTER et al. 1977a). The long-term bone marrow culture system provides a microenvironment of stromal cells which are essential for the continued maintenance of the hematopoietic stem cells (DEXTER 1982). Retroviruses carrying several different oncogenes have been used to infect these cultures and both effects on the ability of the adherent stromal cells to sustain the stem cells and on the differentiation of the stem cells have been evaluated (reviewed in BOETTIGER and DEXTER 1984a).

It is more difficult to pursue such studies in primary cultures of cells derived from solid tissues due to a poorer understanding and consequent inability to manipulate the factors which govern the differentiation of these stem cells. A number the solid tissue systems described above do undergo a limited amount of differentiation in vitro including the kreatinocytes the lens cells, the pigmented retina epithelial cells, and the chondrogenic progenitors in the lumb bud. However, in these systems, it appears that the cells are already determined to a particular lineage and hence, according to the earlier definition, are already programmed.

4.2 Tumor Cells and Cell Lines

Tumor cells grown in culture and established cell lines often continue to synthesize of many of the major differentiation products of the normal cells from which they have been derived and so may be useful for studies of oncogene effects on differentiation. Since either the neoplastic event which leads to a tumor or the establishment of permanent cell lines requires some pertubation of the control of cell differentiation, it is necessary to use some caution in the application of these systems to processes of cell differentiation (BISSELL 1981). On the other hand, these established cell line systems offer some advantages for the analysis of the interactions of oncogenes with cell differentiation programs. The most useful ones are those which can be induced to express more differentiated functions than they do in their proliferating phase. Systems which have been useful in this context include the HL60 promyelomonocytic leukemia cell line (STUDZINSKI et al. 1985; HOLT et al. 1988), the Friend erythroleukemia derived cell lines (FRIEND et al. 1971), the PC12 pheochromocytoma (BAR SAGI and FERAMISCO 1985; ALEMA et al. 1985b), the myogenic cell lines L6 (YAFFE 1968) and C2C12, and the multipotential cell line C3H 10T1/2 (DAVIS et al. 1987). Cell lines offer the advantages of simplicity of manipulation, uniform response, the ability to clone and thus select genetic variants, and access to cell types which may be present in the animal at very low population levels.

One type of cell line which deserves special consideration are those derived from the long-term bone marrow cultures. Some of these lines, FDCP-1 (DEXTER et al. 1980) and 32D (GREENBERGER et al. 1979b) in particular, are factor dependent and have been important in the analysis of growth factor responses. Several cell lines have been isolated which are multipotential and provide potential models for analysis of lineage programming decisions (GREENBERGER et al. 1983; SPOONCER et al. 1986). One series, isolated long-term bone marrow cultures infected with a retroviral vector carrying a *src* oncogene, retains the ability to produce colony-forming units–hematopoietic (spleen) (CFU–S) in vivo, suggesting that it retains most of its normal cell programming and is very similar to the normal hematopoietic stem cells (BOETTIGER and DEXTER 1984b, 1986; SPOONCER et al. 1984).

4.3 Transgenics

The ability to transfer genes into the germ line and have them expressed during development and in the adult provides another system for testing the interaction of oncogenes with developmental programs. It offers some advantages over the culture systems in that the expression occurs within a normal microenvironment. This removes some of the objections to the artificial conditions imposed by tissue culture and allows for the analysis of cell types for which tissue culture methods which preserve normal functions have not yet been devised. On the negative side, analysis of these systems is more complex and the experimental manipulations more limited.

The majority of the published data on use of oncogenes in the construction of transgenic animals has used either the SV40 viral large T antigen or the c-*myc* proto-oncogene. These oncogenes have been linked to a variety of promoters, insulin (BENDER and PEIFER 1987), protamine 1 (BEHRINGER et al. 1988), major histocompatibility class 1 (REYNOLDS et al. 1988), δ-crystallin (MAHON et al. 1987), immunoglobulin heavy chain (KNIGHT et al. 1988; HARRIS et al. 1988; SUDA et al. 1987), murine mammary tumor virus long terminal repeat (LTR) (LEDER et al. 1986), and whey acidic protein (SCHOENENBERGER et al. 1988; ANDRES et al. 1987). In each of these cases tumors are produced in tissues which are targets for high levels of oncogene expression and each promoter produces a different pattern of tumor types. Since only a small proportion of the target cells which express the oncogene produce the tumors, it appears that oncogene expression is necessary but not sufficient to induce the tumors (QUAIFE et al. 1987; ANDRES et al. 1987). These studies have identified additional cell types which can respond to particular oncogenes and produce tumors. Development appears to proceed normally in most tissues. While some studies demonstrate expression of the oncogene in tissue which do not produce tumors, this aspect is less well documented and will require more attention in order to define the cell types which are susceptible to transformation by particular oncogenes.

Analysis of differentiation function in oncogene transgenics is just beginning. Transgenic mice which contain c-*myc* driven by an immunoglobin enhancer fail to produce a population of resting B cells (LANGDON et al. 1986). This may be the result of perturbation of the differentiation process, as suggested by the authors, or due to an inability of cells expressing high levels of c-*myc* to go into a resting state as suggested by other systems (see below). Transgenic mice carrying the *mos* oncogene linked to a Moloney murine leukemia virus LTR gave the highest level of expression in the eye, where expression resulted in reduced basal lamina synthesis and a distortion of the fiber cells without any indication of neoplastic transformation. Perhaps the most interesting of these models is the transgenics produced by the introduction of c-*fos* driven by a metalothionine promoter (RUTHER et al. 1987; STEWART et al. 1987). These animals show a derangement of bone development which has its onset with the initiation of bone development. This is the same tissue which was the primary target for the FBJ retrovirus in which the *fos* oncogene was first identified (FINKEL et al. 1966).

4.4 *Drosophila*

Another system which shows great promise for the analysis of the developmental functions of the proto-oncogenes is *Drosophila*. *Drosophila* homologues to *abl* (HENKEMEYER et al. 1988), two *src*-related genes (GREGORY et al. 1987), *ets* (PRIBYL et al. 1988), *myb* (KATZEN et al. 1985), *raf* (MARK et al. 1987), *int-1* (RIJSEWIJK et al. 1987; CABRERA et al. 1987; UZVOLGYI et al. 1988), and *ras* (SEGAL and SHILO 1986a) have been identified. The tissue-specific expression of several of these suggests that they play a role in development. *abl*, the two *src* genes, and *ras* are present in maternal mRNA and expressed in embryogenesis; in the larval stages they tend to be expressed preferentially in tissues with high proliferation rates (Lev et al. 1985, 1984; Wadsworth et al. 1985; Segal and Shilo 1986b). In the adult, expression is elevated in ovaries (Segal and Shilo 1986b). One of the *src* genes is expressed at high levels in muscle and neural tissue, suggesting a function other than proliferation (Simon et al. 1985). These distributions and conclusions are similar to those derived from studies on vertebrates (see above). Thus, *Drosophila* may provide a reasonable model for proto-oncogene function in development.

The identification of vertebrate proto-oncogene homologues with *Drosophila* developmental genes provides additional fuel to the hypothesis that these vertebrate proto-oncogenes are indeed developmental regulators. The *int-1* oncogene which was identified by promoter insertion of murine mammary tumor virus has extensive sequence homology with the *Drosophila* gene *wingless* (Rijsewijk et al. 1987), and antisense *int-1* RNA induces a phenocopy of the wingless phenotype (Cabrera et al. 1987). The *wingless* gene is one of the segment polarity class of *Drosophila* development genes and results in the absence of cuticular structures in each of the adult body segments and a reduplication of the remaining structures (Banerjee et al. 1987). Another developmental gene, *sevenless*, bears homology the tyrosine kinase class of proto-oncogenes (Hafen et al. 1987). The product of this gene is required for the development of the photoreceptors in the eye and is probably involved in cell-cell interactions essential for the differentiation of the R7 cell.

The importance of the *Drosophila* system is illustrated in the isolation of *abl* deletion mutants in *Drosophila* (Henkemeyer et al. 1987). The real surprise came with the realization that these mutants had a relatively mild phenotype consisting of some recessive lethality at the pharate adult pupal stage, reduced longevity, reduced fecundity, and an irregular pattern of retinal cells. This suggests that the defect may be compensated by other tyrosine kinases. If this is the correct interpretation, then it will be exceedingly difficult to get a proper handle on the normal function of these proto-oncogenes in the absence of a well-developed genetic system such as *Drosophila*.

5 Developmental Classification of Oncogenes

There have been a number of attempts to classify oncogenes on the basis of protein sequence comparisons. The following is an attempt to classify oncogenes on the basis their potential for interaction with particular signalling pathways and their possible mechanism for perturbing cell differentiation. Of necessity, part of the classification is based on rather speculative arguments, but it is hoped that it will provide a useful context for thinking about the potential developmental interactions.

5.1 Peptide Hormones and Their Receptors

The best understood class of oncogenes are those which represent peptide hormones or growth factors and the cellular receptors for these growth factors. The peptide hormone binds to a specific cellular receptor on the external face of the plasma membrane, which results in the activation of a tyrosine kinase on the cytoplasmic face of the membrane. Signals initiated by the receptor-ligand binding appear to be transmitted via the activated kinase. The specificity of the system is determined by the secretion and availability of a specific peptide hormone, and the expression pattern of the receptors on particular cell types and stages of development. This notion is supported by the transfection of the human c-*fms* gene (receptor for the monocyte growth factor CSF-1; SHERR et al. 1985) into NIH 3T3 cells (ROUSSEL et al. 1987). The transfected cells can now be induced to proliferate by human CSF-1, to which mouse cells do not normally respond. This suggests that the intracellular elements in the signalling pathway are shared between the two cell types and only need to be phosphorylated by the activated tyrosine kinase of this exogenous receptor. This mode of action may also explain the nature of the transformation event elicited by the peptide hormone receptor derived oncogenes. In the formation of the oncogenic form the kinase activity is uncoupled from the ligand binding such that it is expressed constitutively (SACCA et al. 1986; DOWNING et al. 1988). Thus the simple model is that the cell is now continuously receiving signals to divide via its normal intracellular signal pathways. One other prediction of this model is that both the cells which normally express a particular receptor and those which do not express this receptor but which contain the intercellular links in the pathway should be transformable by the oncogenic form of the receptor. This applies to the *erbB* EGF receptor oncogene (VELU et al. 1987; LAX et al. 1985; BEUG et al. 1979; CARPENTER, 1987), and the *fms* CSF-1 receptor oncogene (WHEELER et al. 1986; DOWNING et al. 1988; HEARD et al. 1987a; WHEELER et al. 1987).

The case of the growth factor oncogene involves a short-circuiting of the normal regulatory pathways. Usually the same cell does not respond to the growth factor which it is able to produce. This preserves the ability to regulate the system. However, when a cell can both produce and respond to the factor, then there is autocrine stimulation of cell proliferation. In this case it would be predicted that no mutations in the coding region of the factor, but only its expression in inappropriate cells, would induce the oncogenic process. This prediction has proved correct both in the case of the *sis* oncogene which is the homologue to the β-subunit of PDGF (DOOLITTLE et al. 1983; GAZIT et al. 1984; CLARKE et al. 1984) and for the conversion of IL-3 to an oncogene by its artificial transduction to a retroviral vector (HAPEL et al. 1987).

Thus the result of action of either of these forms of the oncogene is to raise the tyrosine kinase activity of the receptor and initiate a signal process to evoke the preprogrammed response of induction to proliferation. However, these peptide hormone-receptor systems also affect processes of cell differentiation. Our current perspective is distorted because the primary assays used are those which require and measure cell proliferation. The hematopoietic growth factors appear to influence not only the proliferation but also and the differentiation and survival of the target cells (Tushinski et al. 1982; Metcalf 1977; Dexter 1984; Ihle 1986; Dexter et al. 1980). It is not clear how these three activities are related or whether there are three separate effects of these hormones. Thus, the separation of cell proliferation stimuli and cell differentiation stimuli is not easily achieved. The hematopoietic growth factors can drive the proliferation of cell lines in the absence of cell differentiation but this is probably not the general case in vivo. Actually, the interactions of these growth factor and growth factor receptor gene families may extend to primary effects on cell differentiation and direct cell-cell interactions, as suggested in the following section.

Several other oncogenes appear to be members of this group, including the tyrosine kinase receptor genes for as yet unknown ligands *neu* (Yamamoto et al. 1986), *kit* (Besmer et al. 1986; Yarden et al. 1987), *met* (Dean et al. 1985), and possibly *ros* (Neckameyer et al. 1986); and additional factor-related oncogenes *int-2* (Dickson and Peters 1987) and *hst* (Delli Bovi et al. 1987; Taira et al. 1987).

5.2 Cell Interaction Molecules

Included in this category are genes which are involved in cell-cell interaction or communication and function on the external face of the plasma membrane but which are functionally distinct from the protein hormone – tyrosine kinase receptor. It probably contains genes with quite distinct modes of action, but I have tentatively grouped them together primarily for lack of sufficient information. There are at least three subgroups represented:

1. Diffusible or semidiffusible proteins which interact with specific cell receptors but which do not elicit a tyrosine kinase activity as a result of the ligand binding. These include the fibroblast growth factor (FGF) related oncogene *int-2* (Dickson and Peters 1987), and one isolated from both Kaposi's sarcoma (Delli Bovi et al. 1987) and from a stomach tumor (*hst*) (Taira et al. 1987); and *int-1*, which is homologous to the segment polarity gene in *Drosophila*, *wingless* (Rijsewijk et al. 1987).
2. Tyrosine kinase transmembrane proteins involved in direct cell-cell interaction rather than interaction with hormone ligands, as represented by the *sevenless* gene in *Drosophila* or perhaps the *ros* oncogene (Hafen et al. 1987).
3. Transmembrane proteins which have growth factor or growth factor-like repeats in their extracellular domain, like *lin-12* in *C. elegans* and notch in *Drosophila* which contain multiple EGF repeats (Greenwald et al. 1983; Wharton et al. 1985). The available data suggests that these genes function in the cell-cell or cell-matrix interactions which are essential to developmental regulation.

FGF contains a heparin-binding domain and is found associated with the extracellular matrix in a precursor form which requires proteolytic activation (Gospodarowicz et al. 1987a). Its effect on cells can be either the stimulation of cell proliferation, as in mesodermally derived cells, or induction differentiation in a variety of cell types including chondrocytes, endothelial cells, and neuronal cells (Gospodarowicz et al. 1987a, b). FGF acts to stimulate myogenic cell growth and hence delays differentiation (Spizz et al. 1987). In muscle tissue FGF reserves are present bound in the heparin sulfate of the extracellular matrix, perhaps for a potential role in tissue repair (KARDAMI et al. 1988). Thus, it is another component of the extracellular matrix which could function either in regulation of muscle differentiation or repair. It has been suggested that FGF may act on differentiation through its effects on the synthesis of extracellular matrix components (GOSPODAROWICZ et al. 1987b). This action may intersect with the action of some of the tyrosine kinase oncogenes in the *src* family (see below). The members of the FGF family which have been identified as oncogenes encode relatively long polypeptides in comparison to other growth factors (DICKSON and PETERS 1987; DELLI BOVI et al. 1987). The added sequences may serve to attach them to the matrix as in the case of FGF or to associate them with the cell membrane as found for CSF-1 (HEARD et al. 1987b). It has been suggested that *int-1* shares structural similarities with the growth factors (BENDER and PFEIFER 1987), in spite of the very restricted tissue distribution of its proto-oncogene homologue in both mammals and *Drosophila* (WILKINSON et al. 1987; SHACKLEFORD and VARMUS 1987; RIJSEWIJK et al. 1987; BENDER and PEIFER 1987). Its homologue in *Drosophila* is required for the proper expression of the structures in each segement; without it certain structures, including the wings, are deleted and replaced by duplicates of other structures (BENDER and PEIFER 1987). The developmental limitations of *int-1* action are evidenced in its ability to transform cells derived from mammary epithelium by not fibroblast-like cells (BROWN et al. 1986).

The *sevenless* gene of *Drosophila* encodes a protein with a large extracellular comain, a transmembrane domain, and a region of homology to the tyrosine kinase oncogenes, and in particular the *ros* oncogene (HAFEN et al. 1987). This gene is required for the normal development of the R7 cell of the eight-cell photoreceptor system in the eye (BANERJEE et al. 1987). The *lin-12* gene in *C. elegans* (GREENWALD 1987) and the *Notch* gene in *Drosophila* (WHARTON et al. 1985) both encode transmembrane proteins which contain multiple EGF repeats in their external domain and are probably directly involved in cell-cell interactions which determine cell fates in development. *lin-12* governs binary decisions between alternative cell fates at several points in development (GREENWALD et al. 1983). The *Notch* function is required for a decision by multipotential precursor cells between alternative neural or epidermal cell fates (LEHMANN et al. 1983).

There may be a role for this class of genes in early embryonic induction. FGF and transforming growth factor B (TGF-B) appear to act in synergy in the induction of mesoderm in early *Xenopus* development (KIMELMAN and KIRSCHNER 1987; WEEKS and MELTON 1987). There is also the possibility that EGF-like molecules could be involved in early embryonic events. The lineage determination mutant *lin-12* in *C. elegans* (GREENWALD et al. 1983), and the *Notch* mutant in *Drosophila* (WHARTON et al. 1985) contain a series of EGF-like repeats. The presence of the EGF receptor has been reported in teratocarcinoma cells (ADAMSON 1987). These intriguing results suggest the existence of a variation on the common peptide hormone–receptor interaction models which may use elements of the same recognition systems but which are not primarily involved in cell proliferation.

5.3 Membrane-Associated Tyrosine Kinase

A large number of oncogenes encode a tyrosine kinase and so show regions of sequence homology on that basis. Sometimes this is referred to as the *src* family of oncogenes. However, on the basis of other structural features which are essential for the function of these genes they appear to be involved in different cell signalling/ recognition systems. This section is concerned with *src* and closely related genes. The v-*src* oncogene is localized to the adhesion plaques of cells (Rohrschneider et al. 1983). It appears to require both a membrane association, as demonstrated by its requirement for myristilation (Kamps et al. 1985; Cross et al. 1984) and a localization to a Triton-insoluble fraction for its proper function in cell transformation (Hamaguchi and Hanafusa 1987; Linder and Burr 1988). c-*src* has not been as thoroughly investigated but its structure should give it a similar subcellular distribution. Thus, it appears that its position in the cell signalling process requires this localization. Much less information is available about the other members of the *src* family, but their structure and phosphorylation targets (Kamps and Sefton 1988) suggest a similar strategy of activity for *yes* (Kitamura et al. 1982), *fgr* (Naharro et al. 1983; Nishizawa et al. 1986), *lyn* (Yamanashi et al. 1987), *syn* (Semba et al. 1986), and *lck* (Voronova and Sefton 1986). The requirement for a membrane localization and the absence of an extracellular domain distinguishes this class from the growth factor receptor oncogenes and from other tyrosine kinase oncogenes, which do not appear to have the same localization requirements.

The association of these protein kinases with the adhesion plaques suggests that they may play a role in the regulation of cell adhesion. Indeed, the *src* oncogene phosphorylates a number of the proteins associated with adhesion plaques, including vinculin (Sefton et al. 1981), talin (DeClue and Martin 1987; Pasquale et al. 1986), and integrin (Hirst et al. 1986), and possibly additional adhesion plaque components (Linder and Burr, 1988). Adhesion plaques are not only the structures through which cells are attached to the culture dish, but they also represent junctional structures through which the cytoskeleton of the cell is connected with the extracellular matrix (Kleiman et al. 1981). The attachments are not only important in determining cell position through cell migration (Ali and Hynes 1978), and in determining cell morphology, but are used in the delivering of developmental signals (Menko and Boettiger 1987).

While adhesion plaques may represent a major "junctional" structure in mesenchymal cells, epithelial cells have several additional junction types. Gap junctions are involved in some cell-cell communication and possibly also in the passing of developmental signals (Warner et al. 1984; Fraser et al. 1987). In lens epithelial cells the expression of the *src* oncogene causes a disruption of these junctions (Menko and Boettiger 1988; Azarnia et al. 1988). It is apparent that this occurs at a structural level since the rate of synthesis of the major junctional protein MP28 is not altered. This suggests that this class of oncogenes may be involved in the regulation or at least the disruption of other types of junctions and hence exert a major influence on cell communication.

By the present classification, the tyrosine kinase oncogenes are divided into several classes, based on the signalling mechanisms with which they may interact. This raises the question of crossover from one system to another by an overactive kinase produced by the activation of a tyrosine kinase proto-oncogene to full oncogenic potential. On the basis of the pattern of substrates which are phosphorylated, v-*src* and v-*yes* show extensive overlap with v-*fps* and slightly less overlap with v-*erbB* and v-*ros* (Kamps and Sefton 1988). This suggests that v-*fps*, v-*erbB* and v-*ros* could be phosphorylating the same critical substrates as v-*src*. However, the absence of v-*ros* from adhesion areas as determined by detergent solubility suggests that it is other substrates which are critical for transformation (HAMAGUCHI and HANAFUSA 1987; JOVE et al. 1986). v-*fms* a member of the tyrosine kinase receptor class, shows very little if any phosphorylation of these substrates (SHERR 1987), and the data suggest that it functions more as a constitutive CSF-1 receptor (SHERR 1987). On the other hand, *erbB*, another member of the tyrosine kinase receptor class, has considerable overlap with v-*src*, suggesting that it may transform by phosphorylation of v-*src* targets rather than acting as a constitutive EGF receptor and interacting with the normal substrates of that pathway. One of the more interesting aspects of this crossover process may be seen in the conversion of factor-dependent hematopoietic cell lines to factor independence (WATSON et al. 1987; PIERCE et al. 1985). The factor independence is accompanied by the development of tumorigenicity (PIERCE et al. 1985). Factor independence can be produced by infection with a retroviral vector carrying the v-*src* oncogene (OVERELL et al. 1987), suggesting that members of this class may be capable of delivering a constitutive growth factor signal in certain contexts. Thus it appears likely that there will be some substrate crossover, which will make it more difficult to determine which of the signalling pathways is involved in a particular transformation event.

5.4 *ras*

The *ras* family of oncogenes consists of three members, k-*ras*, H-*ras*, and N-*ras* (LOWY and WILLUMSEN 1986). Each encodes a protein which binds GTP and hydrolyzes it to GDP. *ras* mutants which decrease the rate of hydrolysis induce cell transformation and are oncogenic (WILLUMSEN et al. 1986). The proteins encoded by *ras* are thought to act on regulatory subunits of some yet to be identified enzyme(s).

In the developmental context the *ras* oncogene effects appear to parallel those of *src*, suggesting that they may function in the same signalling or signal response pathway. Microinjection of antibody to *ras* appears to reverse *src*-induced cell transformation, suggesting that a functional *ras* product is required (SMITH et al. 1986). In differentiating systems both *src* and *ras* block myogenic differentiation in a similar way (FISZMAN and FUCHS 1975; MENKO and BOETTIGER 1987; OLSON et al. 1987; PAYNE et al. 1987), and both induce the extension of neurite processes by PC12 cells (ALEMA et al. 1985b; BAR-SAGI and FERAMISCO 1985). While *ras* has been less exensively tested on different cell types than *src*, a similar suppression of differentiation pattern emerges from these studies (WEISSMAN and AARONSON 1985; NAKAGAWA et al. 1987; BUICK et al. 1987). In the hematopoietic system both *src* and *ras* induce a delay or cause a partial suppression of erythrocyte differentiation (HANKINS and SCOLNICK 1981; KAHN et al. 1986). The only documented difference is in the conversion of factor-dependent lines to factor independence, which is accomplished by the *src* oncogene but not by the *ras* oncogene (WATSON et al. 1987; REIN et al. 1985). Perhaps this difference relates to the possible crossover of *src* into the growth factor receptor pathway discussed above.

5.5 Cytoplasmic Kinases

Many of the membrane-associated signalling events associated with tyrosine kinase receptors, transformation by *src*, and activation of the protein kinase C and diacylglycerol pathways result in the phosphorylation of a number of cellular proteins on threonine and serine as well as tyrosine (MACARA 1985). Some of these phosphorylations are secondary and suggest the possibility of a kinase cascade as an element of the signal transmission and cellular response to these membrane signals. Two classes of cytoplasmic kinases may be recognized among the oncogenes: the tyrosine kinases and the serine-threonine kinases.

Among the tyrosine kinases the *abl* and the *fps/fes* oncogenes can function like the members of the *src* family described above; however, there are indications that they may be derived from genes with a primarily cytoplasmic location and thus may play some distinct roles in cellular regulation. Both *fps* and *abl* are activated in part by the addition of viral *gag* sequences (FOSTER et al. 1985; PRYWES et al. 1983) which would be expected to direct them to the membrane since that is part of the *gag* function in virus replication. The temperature-sensitive mutant of *fps* appears to be temperature-sensitive for membrane association and appears to require the membrane location for expression of its transforming function (MOSS et al. 1984). In the case of c-*abl*, there is an alternative splicing pattern which gives different N-termini to the protein (BEN-NERIAH et al. 1986) and potentially different subcellular addresses. While there is no direct support for the possibility that cytoplasmic versions of these oncogenes can transform, it would be interesting to test this possibility, particularly in cells which normally express high levels of the homologous proto-oncogene.

The *mil/raf* and *mos* oncogenes are generally considered to be cytoplasmic in location, although evidence for this is limited (PAPKOFF et al. 1983). They share significant homology with the tyrosine kinase oncogenes but the kinase activity displayed appears to be specific for serine and threonine rather than tyrosine (SETH et al. 1987; MOELLING et al. 1984). In addition to their kinase activity, both encode gene products which will bind DNA and RNA in vitro (SETH et al. 1987; BUNTE et al. 1983), although no sequence specificity of this binding has been described. It is possible that this binding activity is responsible for the suppression of collagen promoter by the *mos* oncogene demonstrated in transfection assays (SCHMIDT et al. 1985), or for the transactivation of α-fetoprotein by *raf* as demonstrated in vitro (VOGT et al. 1987). It remains to be determined whether these effects occur in vivo and are due to a direct and specific effect of the oncogene, but they do suggest that these products of these oncogenes may function in the nucleus rather than in the cytoplasm. In terms of their kinase activity, little is known of potential targets. In the context of differentiating systems, both oncogenes suppress the differentiated phenotype of several cell types. Infection of thyroid epithelial cells with a *mos* retrovirus resulted in the loss of thyroglobulin secretion, loss of iodide uptake, and loss of dependence on six growth factors including thyrotropin (FUSCO et al. 1985). In fibroblasts *mos* suppresses the synthesis of both fibronectin and type I collagen (SETOYAMA et al. 1985), which parallels the effect of *src* on fibroblasts. The *mil/raf* oncogene appears to be a member of a larger family including *raf-2* (HUEBNER et al. 1986) and *pks* (MARK et al. 1986).

5.6 Nuclear Transcription Complex

The nuclear oncogenes *jun* (MAKI et al. 1987), *fos* (CURRAN and TEICH 1982), and *erbA* (WEINBERGER et al. 1947; SAP et al. 1986) have been directly implicated in gene activation. *jun* appears to be the homologue of the human transcription factor AP-1 (BOHMANN et al. 1987) and closely related to the yeast transcription factor GCN4 (VOGT et al. 1987). The c-*fos* gene product has been found specifically associated with transcription complexes on the *aP2* gene, a differentiation marker in 3T3-derived adipocytes, following induction of adipocyte differenatiation (DISTEL et al. 1987). The *erbA* oncogene is the homologue of the thyroid hormone receptor and is thus involved in the binding to hormone-responsive elements in the chromosome (WEINBERGER et al. 1987; SAP et al. 1986). DNA binding activity has also been reported for the *myc* (WATT et al. 1985; BUNTE et al. 1983; BUNTE et al. 1984) and *myb* (MOELLING et al. 1985) oncogenes, though no sequence specificity has been described. Many hypotheses have been put forward on the function of *myc*, but most have not been substantiated. It is not associated exclusively with cell proliferation since it is expressed in many nonproliferating cells and may even increase in cells as they withdraw from the cell cycle (ENDO and NADAL GINARD 1986). It is not one of the factors associated with DNA polymerase (STUDZINSKI et al. 1986). It does not appear to be associated with splicosomes (SULLIVAN et al. 1986). The *myb* oncogene has been the subject of fewer hypotheses, but it does have a clear association with hematopoietic cells and that alone suggests developmental regulation and particular function in these cells. More interestingly is its expression in early stages of differentiation (see above) and the dominant nature of its transforming action on differentiated macrophages (Ness et al. 1987). Given the associations of *erbA* and *jun* directly with transcriptional control and the implication of specific association of *fos* with the transcriptional control of a developmental marker, it is tempting to speculate than both *myc* and *myb* as well as the less intensively studied oncogenes *rel* (BROWNELL et al. 1987) and *ski* (LI et al. 1986) will be involved at some level in transcriptional control.

The emerging picture of communication between a cell and its environment has developed several consistent themes. Cellular receptors interact with extracellular ligands; this may turn on a tyrosine kinase, activate *ras*, activate the diacylglyerol phosphorylation pathway leading to the generation of second messengers, alter ion flux, and/or activate protein kinase C. The actual wiring of these pathways is complex and still poorly understood; however, this activity at the membrane is often accompanied by the increases in mRNA levels for the *myc* and *fos* and occasionally the *myb* proto-oncogenes. Elements of this general pattern, including the rapid induction of *myc* and/or *fos*, have been reported for stimulation of T cells by IL-1a (KOVACS et al. 1986), IL-2 (REED et al. 1985), or antigen (LACY et al. 1986); stimulation of B cells by BCGF (LACY et al. 1986), TPA (NILSSON et al. 1986), or antigen (KAUFMANN et al. 1987); stimulation of fibroblasts by EGF (BRAVO et al. 1985; RAN et al. 1986) or PDGF (WOME et al. 1987; KRUIJER et al. 1984); stimulation of thyroid cells by TSH (DERE et al. 1985); and stimulation of neural cells with NGF (MILBRANDT 1986; KRUIJER et al. 1985; CURRAN and MORGAN 1985). This is not an exhaustive list and in any case probably only represents the tip of the iceberg. The pattern is clear and suggests that the gene products of *myc* and *fos* are required for a proper response of a primed cell to a specific stimulus. This implies that the specificity of these responses is governed by the expression of a particular receptor by the cell and by the cell program, which responds by activation or inactivation of different specific gene sets in response to the actions of *myc* and *fos* proteins.

Based on gene transfer using retroviruses or retroviral vectors carrying either the *myc* or *myb* oncogene or proto-oncogene, there have been attempts to classify these genes in terms of the type of effect they have on the expression of different developmental programs. The model in the case of *myc* is that it affects the cells ability to withdraw from the cell cycle but has no direct effect on the expression of differentiation markers. Thus, in lineages such as the chondroblast (ALEMA et al. 1985a) or macrophage (DURBAN and BOETTIGER 1981a; NESS et al. 1987) there is no or very little effect of *myc* expression on the developmental markers because expression of these markers occurs in dividing cells. On the other hand, in cell systems in which withdrawal from the cell cycle is essential for terminal differentiation such as the myoblast (ENDO and NADAL GINARD 1986; FALCONE et al. 1985) or erythroblast (COPPOLA and COLE 1986; DMITROVSKY et al. 1986a, b), *myc* expression could block differentiation. Considering the time requirement for *myb* switch-off in erythrogenesis, one must interpret the effect of *myc* as being on the commitment to terminal differentiation which still leaves a fixed number of divisions. In a transgenic mouse carrying a *myc* gene on an immunoglobulin promoter, the major developmental defect was the absence of resting B cells (LANGDON et al. 1986). One interpretation of these data (though not the one given) is that the constitutive expression of *myc* prevented withdrawal from the cell cycle. These notions would be much more satisfying if we had more information as to the nature of the signals which trigger withdrawal from the cell cycle as opposed to induction of differentiation. The model for *myb* is that its function is to suppress the expression of terminal cell markers and thus allow sufficient expansion of the unipotential progenitors. This is discussed in detail below.

6 Model Systems

The most direct approach to analysis of the effects of oncogenes on developmental programs is through gene transfer experiments. The presence of many of the oncogenes in retroviruses has made this task easier, though newly constructed retroviral vectors with encoded oncogenes have also contributed to these studies. The discussion is largely restricted to primary culture systems and concentrates on those for which the most data are available as paradigms for the less-developed systems.

6.1 Hematopoiesis

Two general approaches have been used in hematopoietic culture systems for analysis of effects of oncogenes on cell differentiation: (1) the long-term bone marrow culture system which maintains the hematopoietic stem cells and permits their continuous differentiation, and (2) the colony and in vitro culture systems which are permissive for individual hematopoietic lineages.

The long-term bone marrow culture systems maintains the hematopoietic stem cells in culture for periods of a year or more (DEXTER et al. 1977a; DEXTER 1982; BOETTIGER and DEXTER 1984a). The stem cells can be assayed by the spleen colony assay (TILL and MCCULLOCH 1961) or by the reconstitution of lethally irradiated recipients (BOETTIGER and DEXTER 1986). The latter is a particularly stringent test of normality since the stem cell must repopulate all the hematopoietic lineages. The maintenance of the stem cells is accomplished by the inclusion of cells which produce an adherent, heterogeneous monolayer in culture (ALLEN and DEXTER 1976; DEXTER 1982) which is representative of the normal bone marrow microenvironment. When these cultures are infected by retroviruses carrying specific oncogenes, both the adherent cells and the hematopoietic cells may be infected. With viruses caryying either the *mos*, *abl*, or *ras* oncogenes there is a rapid transformation of the monolayer, resulting in hematopoietic failure and loss of the stem cells (DEXTER et al. 1977b; TEICH and DEXTER 1978). Unexpectedly, the effect of *src* on these cultures was quite different. The hematopoietic cells derived from the *src*-infected cultures showed a decrease in the proportion of differentiated cells (mostly granulocytes in these cultures in the absence of erythropoietin) and a >10-fold increase in the more immature cells which was apparent either in the composition of nonadherent cells in the original cultures or in agar colonies derived from these cells (BOETTIGER et al. 1984). These immature cells were multipotential as determined by the production of spleen colonies (BOETTIGER and DEXTER 1984b, 1986), ability to reconstitute lethally irradiated animals (BOETTIGER and DEXTER 1986), production of multilineage mixed colonies in agar suspension (SPOONCER et al. 1984), and the ability to establish multipotential cell lines from these cultures (SPOONCER et al. 1984, 1986). While this was a relatively stable change in the multipotential hematopoietic cells, as evidenced by its persistence in other environments for at least 10 passages (BOETTIGER et al. 1984; SPOONCER et al. 1984), the effect of *src* was indirect, since *src* was not expressed in the altered hematopoietic cells (WYKE et al. 1986). Given the effects of *src* expression on extracellular matrix synthesis, it now appears most likely that the effect on the hematopoietic stem cells was mediated by such changes in the hematopoietic microenvironment. Thus, while these experiments do not provide evidence that the hematopoietic stem cells can be directly altered by expression of oncogenes, this remains the only system in which their behavior has been affected by an oncogene. This system holds great potential for the analyses of the effects of other oncogenes, but the necessary manipulations and the fine tuning of the system to allow the desired analysis are not simple to achieve.

Short-term assays have been employed using cells derived from either bone marrow or, for embryonic stages, murine fetal liver or chicken yolk sac cells. There appears to be no effect of the oncogene-carrying retroviruses on the transformation of cells in the granulocyte lineage, so the studies have focussed on the erythroid and the lymphoid lineages. Bone marrow-derived erythroid cells have been infected with naturally occurring retrovirus carrying the oncogenes *erbB* (FRYKBERG et al. 1983), *erbA* (KAHN et al. 1986), *erbB* + *erbA* (BEUG et al. 1979; SAMARUT and GAZZOLO

1982), *myb* (Beug et al. 1979; Durban and Boettiger 1981b; Gazzolo et al. 1979), *myb + ets* (Radke et al. 1982; Moscovici et al. 1983), *src* (Durban and Boettiger 1981a; Adkins et al. 1984), and *ras* (Holmes et al. 1986). Similar experiments have been performed with mouse fetal liver cells using the *abl* (Waneck and Rosenberg 1981; Moscovici et al. 1983) and *ras* (Hankins and Scolnick 1981) oncogenes. *erbA* alone and *myb* alone did not appear to alter the erythroid cells whereas the other combinations produced an erythroid cell transformation and accumulation of blasts (Beug et al. 1979, 1984, 1982a). Both *ras* and *abl* produced erythropoietin-independent colonies (Hankins and Scolnick 1981; Waneck and Rosenberg 1981). It would appear that these oncogenes cause a delay in erythroid development since the affected erythroid cells do differentiate spontaneously in each of these cases, albeit at rather different rates. A careful analysis of the target cells for infection and the phenotype of the transformed cells following infection of avian bone marrow cells with the avian erythroblastosis viruses caryying both the *erbA* and *erbB* oncogenes showed that the burst forming unit erythroid (BFUe, early erythroid progenitor) was the primary target for infection but that the transformed cells resembled the more mature colony forming unit erythroid (CFUe, later committed erythroid cell) (Samarut and Gazzolo 1982). Thus, some degree of maturation was required for the oncogene to exert its effect, supporting the concept of differentiation-dependent transformation (Boettiger and Durban 1980).

Infection and transformation of macrophage-lineage cells derived from either yolk sac or bone marrow have been performed in the chicken system using retroviruses carrying *myb* (Beug et al. 1979; Moscovici 1967; Gazzolo et al. 1979; Durban and Boettiger 1981a, b), *myc* (Beug et al. 1979; Graf et al. 1981; Ness et al. 1987; Gazzolo et al. 1979; Durban and Boettiger 1981a), *myb + ets* (Radke et al. 1982; Moscovici et al. 1983), and *src* (Adkins et al. 1984; Durban and Boettiger 1981a) oncogenes. Neither the *myc* not the *src* oncogene has substantial effects on the differentiation of the cells (Durban and Boettiger 1981a; Lipsich et al. 1984; Ness et al. 1987), although *myc* and possibly *src* may have some effect on cell proliferation (Durban and Boettiger 1981a; Lipsich et al. 1984; Nrss et al. 1987). The *myb*-carrying viruses, on the other hand, do have a strong effect on cell differentiation. Expression of *myb* suppresses the expression of most of the macrophage-specific functions including accumulation of cytoplasmic lipid, lysosomal acid phosphatase, lysozyme, phagocytosis, Fc receptors, C3 receptors, and some cell surface markers recognized by polyclonal antisera (Durban and Boettiger 1981a; Gazzolo et al. 1979; Radke et al. 1982; Beug et al. 1979, 1982b; Beug and Graf 1981). The morphology of the transformed cell resembles that of an early monoblast but it can be distinguished from the early cells by subtle morphological differences (nuclear location and nucleaus-cytoplasm ratios, occasional vacuoles) and specific cell surface markers of the mature macrophage which include the antigen recognized by a monoclonal antibody (Jurdic et al. 1984) and a cell surface ATPase (Durban and Boettiger 1981a). The suppression is coordinate, but the cell differentiation products fall into two classes, those which are affected and those which are not. The suppression of the major macrophage products prevents their accumulation in the cell and the absence of some of these products permits continued cell proliferation. The proliferation rate gives a doubling time of about

3 days either in vivo or in vitro, which is not very different from that expected for a monoblast (Durban and Boettiger 1981 b).

These results on the effect of the *myb* oncogene on macrophage differentiation and the distribution of the c-*myb* proto-oncogene have led to the hypothesis that the function of *myb* is to suppress the expression of the differentiated cell program in a cell already programmed to be a macrophage, in order to allow expansion of the pool of committed cells (Boettiger and Durban 1980; Boettiger and Olsen 1988). Expression of *myb* in macrophage-lineage cells outside the yolk sac or bone marrow microenvironment which controls their expansion leads to an uncontrolled expansion of this cell population. This hypothesis predicts that the *myb* oncogene and its normal homologue will exhibit a similar function. This remains to be tested. While initially it appeared that this function would be limited to the macrophage lineage, the increasing evidence of elevated expression in the erythroid, B-cell and T-cell lineages (Kirsch et al. 1986; Bender and Kuehl 1986; Sheiness and Gardinier 1984b) suggested that a similar model might apply to those lineages. This idea is strengthened by the demonstration of B-cell transformation (tumors) produced by promoter insertion in the c-*myb* gene (Kanter et al. 1988), thus providing an analogue of the transformation of macrophage lineage cells by v-*myb*. Also, induced differentiation of Friend erythroleukemia cells is prevented by elevated c-*myb* expression produced by transfection by a *myb* expression vector (Clarke et al. 1988b).

6.2 Myogenesis

Cultures of cells from solid tissues have formed the basis of much of the tissue culture experimentation with oncogenic retroviruses. Several differentiated cell systems have been developed, as described earlier. In general, these systems have focused on the production of cultures which contain a single cell type or at least a defined and limited number of cell types. These systems are excellent for the analysis of the perturbation of developmental programs which are already expressed or at least primed for expression. Of these culture systems, the most extensively used is the myogenic differentiation. Most of the oncogene work has used primary chick breast muscle-derived cells, though some workers have used myogenic rodent cell lines. In this culture system the mature cells are post-mitotic and easily recognizable since they are multinucleate and will contract spontaneously. This postmitotic phase provides an excellent system for the testing of the effects of oncogenes on expression of differentiated cell products in the absence of cell division. While it has not been extensively used for this purpose, it has been shown that v-*src* can switch off most of the differentiated cell products in the absence of cell division (West and Boettiger 1982). There is the complicating effect that expression of some oncogenes in these postmitotic cells can lead to degeneration (Holtzer et al. 1975). The myogenic system has generally been used at an earlier stage at which a variety of agents can prevent the appearance of most of the muscle-specific products.

Myogenic differentiation offers a unique advantage over other in vitro differentiation systems in that it has been possible to define a specific control point which requires the cells to make a binary decision: to withdraw from the cell cycle and initiate muscle biochemical differentiation or to go into another cell cycle (HOLTZER et al. 1983; ENDO and NADAL GINARD 1987). While there is considerable support for this idea from studies with inhibitors of myogenesis and analysis of myogenesis in culture, the most direct evidence comes from two experiments. In the first, myoblasts transformed with a temperature-sensitive *src* mutant were allowed to proliferate at a permissive temperature or shifted to nonpermissive temperature in the presence or absence of mitomycin C (FALCONE et al. 1984). At the permissive temperature 35% of the cells were labelled with ^{3}H-TdR in 24 hours, whereas at non-permissive temperature 90% of the mitomycin C treated cells afferentiated into multinucleatic myotubes. Thus, the majority of the cells in the original culture had two options and chose which option on the basis of the level of expression of the *src* oncogene. The second experiment used a monoclonal antibody to the fibronectin receptor and demonstrated that blocking of this receptor prevented myogenesis (MENKO and BOETTIGER 1987). This demonstrated that a specific interaction, in this case between the receptor and the substrate, was required for myogenesis to proceed beyond this specific point. The point in myogenesis at which this block occurs is the same in both cases, and in the cases of most of the chemical inhibitors of myogenesis. While it is likely that there are specific control points in other lineage, these have not been defined. The advantage in myogenesis is that the control point coincides with the withdrawal from the cell cycle, which makes it easy to measure precisely.

Muscle-specific genes fall into two classes, those which are synthesized prior to the control point, and those synthesized only after the control point, i.e., following withdrawal from the cell cycle. In the former category are those for desmin (HILL et al. 1986; MENKO and BOETTIGER 1987), the major intermediate filament protein specific to muscle cells, and two monoclonal antibodies, L4 which identifies a cell surface component (GEORGE-WEINSTEIN et al. 1987) and 13F4 which identifies an internal protein (RONG et al. 1987). L4 appears to be the earliest detectable marker and is particularly useful in distinguishing myogenic and nonmyogenic cells which are not distinguishable morphologically in early cultures. The latter category includes the major muscle-specific proteins, the muscle myosin, α-actin, muscle tropomyosins, acetylcholine, and acetylcholine receptors (HOLTZER et al. 1975; FISZMAN and FUCHS 1975; CHI et al. 1975; NADAL GINARD 1978). The synthesis of these products appears to be controlled at the transcriptional level since the mRNAs for these products cannot be detected prior to withdrawal from the cell cycle (ENDO and NADAL GINARD 1986; NADAL GINARD 1978). Proteins necessary for the initiation of cell fusion are also presumbly included in this general switch-on of muscle-specific products, also referred to as biochemical differentiation, as fusion to myotubes follows as part of this postmitosis phase. While the studies employing oncogenes have not been carried much past the control point, the process of myogenic differentiation continues in the multinucleate myotubes with the process of enevation by nerves and differentiation into fast- and slow-twitch fibers, which requires expression of specific muscle myosins (STARON and PETTE 1987).

Myogenesis has been employed to test the action of the *src* (HOLTZER et al. 1975; FISZMAN and FUCHS 1975; FALCONE et al. 1984), *fps* (FALCONE et al. 1985), *erbA* (FALCONE et al. 1985), *erbA*+*erbB* (FALCONE et al. 1985), *ras* (OLSON et al. 1987), and *myc* (FALCONE et al. 1985) oncogenes. Much of this work has been recently reviewed (ALEMA and TATO 1987). On the basis of failure to fuse to make myotubes and by analysis of the presence of the differentiation markers described above, all these oncogenes, with the exception of *erbA*, appear to suppress expression of postcontrol point markers and prevent the appearance of myotubes. Does this mean that all these oncogenes interact with this developmental program in the same way? *src*, *fps*, and *erbB* are all tyrosine kinases and have the potential of acting like *src*, and *ras* may be part of the *src* signalling/response mechanism (see above). *myc* on the other hand may be functioning quite differently. There are two elements which determine the expression of biochemical differentiation: the decision at the control point and the withdrawal from the cell cycle. In the chondroblast system these two factors are unlinked; *src* suppresses the expression of the chondrogenic markers but *myc* does not (ALEMA et al. 1985a). This raises the possibility that *myc* functions to prevent withdrawal from the cell cycle. The reactivation of *src* in mature myotubes causes a reduction in the rates of synthesis of the muscle-specific products (WEST and BOETTIGER 1982) and over several days a vacuolation and degeneration of the myotubes (HOLTZER et al. 1975). This suggests that it is not only the decision point which can be affected by *src* and raises the question of whether some matrix-derived signal, with which *src* may interfere, needs to be continuously present to maintain the stability of the differentiated phenotype.

The effect of *src* on the differentiation of several other mesenchymally derived cell types has also been examined. These include chondroblasts (PACIFICI et al. 1977; ADAMS et al. 1982; ALLEBACH et al. 1985; ALEMA et al. 1985a), muscle fibroblasts (ALLEBACH et al. 1985), tendon fibroblasts (SCHWARTZ et al. 1978), and adipocytes (LANGER-SAFER et al. 1985). These systems differ from muscle differentiation in the absence of a defined control point. In each of these systems there is a suppression of expression of differentiated cell markers. In the chondroblast and fibroblast systems this suppression is reversible and differentiation is reexpressed following a shift to the nonpermissive temperature (PACIFICI et al. 1977; ADAMS et al. 1982). The expression of these differentiation products is controlled primarily at the level of mRNA, as determined by in vitro translation and Northern blots (ADAMS et al. 1982, 1985; FOCHT et al. 1987; ALLEBACH et al. 1985). Again, this raises the question of whether the stability of the differentiated phenotype requires constant "signals" from the extracellular matrix which are disrupted by *src* or whether *src* is initiating its own signal to unbalance the differentiated state.

6.3 Epithelium

The final set of models involve epithelial tissues. They are included separately because it is likely that the regulation of their differentiated programs differs in some aspects from that of the mesenchymal tissues due to the presence of new cell junctional types and to their particular association with basal lamina. The cell types that have been used include neural retina (CASALBORE et al. 1987;

BECHADE et al. 1985; CALOTHY et al. 1980; NOTTER et al. 1987), pigmented retina (BOETTIGER et al. 1977), lens (MENKO and BOETTIGER 1988), keratinocytes (YUSPA et al. 1985; WEISSMAN and AARONSON 1985), and thyroid cells (FUSCO et al. 1985; NAKAGAWA et al. 1987). In general, these culture systems are more complex than the ones discussed above.

The neural retina and the lens cultures consist of more than one cell type and few studies have focussed on the changes in expression of specific developmental markers. Where it has been examined in neural retina cultures and in other cultures of neural origin, expression of the *src* oncogene does not appear to shut off several normal neural functions (CRISANTI et al. 1985; GIOTTA and COHN 1981). The main focus of the neural retina system has been on the effects of oncogene expression on cell proliferation (CASALBORE et al. 1987; BECHADE et al. 1985; NOTTER et al. 1987; IBA et al. 1985). In the absence of oncogene expression the cells cease division in culture; however, whether the reason for the continued proliferation is the activation of proliferation or the removal of the normal stop signals which accompany the normal course of differentiation is unclear. It is possible that some of the oncogenes stimulate the proliferation and others promote more rapid loss of the normal cell differentiated phenotype, thus removing the stop signal. The effect of the *myc* oncogene on this system depends on whether or not the cells are plated on a collagen substrate (CASALBORE et al. 1987). This suggests that the loss of normal differentiation in the absence of the collagen substrate is an important component of the stimulation of proliferation.

In the lens and the pigmented retina systems, expression of the *src* oncogene induces altered synthesis of extracellular matrix components (D. BOETTIGER and A. S. MENKO, unpublished observations) and changes in cell structure which affect the differentiated cell function. In the pigmented retina cells, melanosomes are exocytosed (BOETTIGER et al. 1977). In the lens cells, the gap junctions are disrupted (MENKO and BOETTIGER 1988). These events result in altered morphology and loss of differentiated cell function. Unlike the myogenic and chondrogenic systems, the lens major gap junction protein mp28 and δ-crystallin, which are major lens cell differentiated products, continue to be made at similar rates following transformation (MENKO and BOETTIGER 1988). These results suggest that the oncogene affects the control of extracellular matrix synthesis and the interaction of the cells with their extracellular matrix, and may not directly affect the synthesis of other cell type-specific products.

The primary rat keratinocyte culture system is one which permits the expression of keratinocyte differentiated functions in culture. The experimental system and the effects of several oncogene on keratinocyte differentiation parallel the myogenic culture system and the effects of oncogenes in that system. The expression of specific cytokeratins, cornification, and the dependence on EGF are characteristic of the differentiated cells. Expression of the *src*, *fes*, *abl*, *ras*, or *mos* oncogenes prevents the expression of the cytokeratins and cornification and removes the normal hormone dependence of these cells (WEISSMAN and AARONSON 1985).

7 Perspectives and Questions

While there appears to be general agreement that oncogenes may affect the expression of cell differentiation and that their proto-oncogene forebears may function as normal developmental regulators, there are few attempts to relate oncogene functions directly to developmental processes. Much of the focus on oncogene function has been on possible roles in cell proliferation. This emphasis is reasonable considering the greater accessibility of culture systems to measure proliferation as opposed to ones which measure differentiation. It is clear that the processes of cell proliferation and cell differentiation are both elements of the cell's developmental program. In the in vivo situation and with the development of neoplasia, it is likely that the mechanisms which prevent cells from dividing which are largely linked to the expression of cell differentiation will prove to be at least as important in controlling cell proliferation as the peptide hormone growth factors. At some stage the complexities of the differentiating system will need to be addressed. It is hoped that the discussion on possible approaches will stimulate research in this area.

The other factor which is often lacking in discussions on oncogenes and development is some common framework for comparison of different approaches and results. Based on our current knowledge of oncogene function and on current and past concepts in developmental biology, I have offered a simplified developmental model which centers on the idea of cell programming and the general functions which a cell must be able to accomplish at different stages of maturity within a cell lineage. Within this context, I have suggested that the majority of the oncogenes function in the social interaction and communication between cells which are essential for the development of a mature animal from a zygote. The oncogenes have been classified on the basis of how they would fit into a variety of intracellular signalling pathways at different levels between the initial interaction and the cell's response. Of necessity, such a scheme is speculative, but may at least provoke the reader to some new viewpoints on potential function. There is an additional hypothesis which is concealed in this view of oncogene function, if one takes the current data on the effects of oncogenes on cell types which already express their terminal differentiation phenotype. The hypothesis is that even these cells require continued signals from their microenvironment in order to continue to express their differentiated cell phenotype. This hypothesis presents a different view on the maintenance of differentiated cell phenotypes, but it is consistent with the loss of expression of the differentiated phenotype with changes in microenvironment brought about by neoplastic transformation or growth under various culture conditions.

The data presented in this review involve both the introduction of oncogenes into various cell types and the expression of proto-oncogenes in various cell types. Both lines of evidence provide information on the cell types in which the particular biochemical pathways used by the oncogenes and proto-oncogenes are operational. As a first approximation, it is assumed that the functions of the oncogene and its proto-oncogene progenitor are similar. Elucidation of these patterns is fundamental to understanding of the particular signals and controls which are necessary for normal development and which may be either usurped or cut in the process of oncogenic transformation. Some of the patterns, particularly those in which oncogene expression increases in terminally differentiated and nondividing cells, will require us to view both processes in a different light.

Acknowledgements. The author is indebted to Drs. A. Sue Menko, Brian Aneskeivich, Beatrice Haimovich, and Mindy George-Weinstein for critical reading of the manuscript. This work is supported by grants from the National Cancer Institute CA16502 and CA30383.

8 References

Adams SL, Boettiger D, Focht RJ, Holtzer H, Pacifici M (1982) Regulation of the synthesis of extracellular matrix components in chondroblasts transformed by a temperature-sensitive mutant of Rous sarcoma virus. Cell 30: 373–384

Adams SL, Pacifici M, Focht RJ, Allebach ES, Boettiger D (1985) Collagen synthesis in virus-transformed cells. Ann NY Acad Sci 460: 202–213

Adamson E (1987) Oncogenes in development. Development 99: 449–471

Adkins B, Leutz A, Graf T (1984) Autocrine growth induced by src-related oncogenes intransformed chicken myeloid cells. Cell 39: 439–445

Alema S, Tato F (1987) Interactions of retroviral oncogenes with the differentiation program of myogenic cells. Adv Cancer Res 49: 1–28

Alema S, Tato F, Boettiger D (1985a) myc and src oncogenes have complementary effects on cell proliferation and expression of specific extracellular matrix components in definitive chondroblasts. Mol Cell Biol 5: 538–544

Alema S, Casalbore P, Agostini E, Tato F (1985b) Differentiation of PC12 phaeochromocytoma cells induced by v-src oncogenes. Nature 316: 557–559

Ali IU, Hynes RO (1978) Effect of LETS glycoprotein on cell motility. Cell 14: 439–446

Allebach ES, Boettiger D, Pacifici M, Adams SL (1985) Control of types I and II collagen and fibronectin gene expression in chondrocytes delineated by viral transformation. Mol Cell Biol 5: 1002–1008

Allen TD, Dexter TM (1976) Cellular interrelationships during in vitro granulopoiesis. Differentiation 6: 191–194

Anderson SJ, Furth ME, Wolff L, Ruscetti SK, Sherr CJ (1982) Monoclonal antibodies to the transformation-specific glycoprotein encoded by the feline retroviral oncogene v-fms. J Virol 44: 696–702

Andres AC, Schonenberger CA, Groner B, Henninghausen L, LeMeur M, Gerlinger P (1987) Ha-ras oncogene expression directed by a milk protein gene promoter: tissue specificity, hormonal regulation, and tumor induction in transgenic mice. Proc Natl Acad Sci USA 84: 1299–1303

Anthony DT, Schuetze SM, Rubin LL (1984) Transformation by Rous sarcoma virus prevents acetylcholine receptor clustering on cultured chicken muscle fibers. Proc Natl Acad Sci USA 81: 2265–2269

Azarnia R, Reddy S, Kmiecik TE, Shalloway D, Loewenstein WR (1988) The cellular src gene product regulates junctional cell-to-cell communication. Science 230: 398–401

Banerjee U, Renfranz PJ, Hinton DR, Rabin BA, Benzer S (1987) The sevenless+ protein is expressed apically in cell membranes of developing Drosophila retina; it is not restricted to the cell R7. Cell 51: 151–158

Bar Sagi D, Feramisco JR (1985) Microinjection of the ras oncogene protein into PC12 cells induces morphological differentiation. Cell 42: 841–848

Barka T, Gubits RM, van der Noen HM (1986) Beta-adrenergic stimulation of c-fos gene expression in the mouse submandibular gland. Mol Cell Biol 6: 2984–2989

Beard JW (1963) Avian virus growth and their etiologic agents. Adv Cancer Res 7: 1–127

Bechade C, Calothy G, Pessac B, Martin P, Coll J, Denhez F, Saule S, Ghysdael J, Stehelin D (1985) Induction of proliferation or transformation of neuroretina cells by the mil and myc viral oncogenes. Nature 316: 559–562

Behringer RR, Peschon JJ, Messing A, Gartside CL, Hauschka SD, Palmiter RD, Brinster RL (1988) Heart and bone tumors in transgenic mice. Proc Natl Acad Sci USA 85: 2648–2652

Bell JC, Jardine K, McBurney MW (1986) Lineage-specific transformation after differentiation of multipotential murine stem cells containing a human oncogene. Mol Cell Biol 6: 617–625

Ben-Neriah Y, Bernards A, Paskind M, Daley GQ (1986) Alternative 5′ exons in c-abl mRNA. Cell 44: 577–586

Bender TP, Kuehl WM (1986) Structure and function of c-myb proto-oncogene mRNA in murine B cells. Curr Top Microbiol Immunol 132: 153–158

Bender W, Pfeifer M (1987) Oncogenes take wing. Cell 50: 519–520

Besmer P, Murphy JE, George PC, Qiu FH, Bergold PJ, Lederman L, Snyder HWJr, Brodeur D, Zuckerman EE, Hardy WD (1986) A new acute transforming feline retrovirus and relationship of its oncogene v-kit with the protein kinase gene family. Nature 320: 415–421

Beug H, Graf T (1981) Production and characterization of antisera specific for the erb-A portion of p75, the presumptive transforming protein of avian erythroblastosis virus. Virology 111: 201–210

Beug H, von Kirchbach A, Doderlein G, Conscience JF, Graf T (1979) Chicken cells transformed by seven strains of defective avian leukemia viruses display three distinct phenotypes of differentiation. Cell 18: 375–390

Beug H, Pamieri S, Freudenstein C, Zentgraf H, Graf T (1982a) Hormone-dependent terminal differentiation in vitro of chicken erythroleukemia cell transformed by ts mutants of avian erythroblastosis virus. Cell 28: 907–919

Beug H, Hayman MJ, Graf T (1982b) Myeloblasts transformed by the avian acute leukemia virus E26 are hormone-dependent for growth and for the expression of aputative myb-containing protein, p135 E26. EMBO J 1: 1069–1073

Beug H, Leutz A, Kahn P, Graf T (1984) Ts mutants of E26 leukemia virus allow transformed myeloblasts, but not erythroblasts or fibroblasts, to differentiate at the nonpermissive temperature. Cell 39: 579–88

Bishop JM (1983) Cellular oncogenes and retroviruses. Annu Rev Biochem 52: 301–354

Bishop JM (1985) Viral oncogenes. Cell 42: 23–38

Bissell MJ (1981) The differentiated state of normal and malignant cells or how to define a "normal" cell in culture. Int Rev Cytol 70: 27–100

Bister K, Nunn M, Moscovici M, Perbal B, Baluda MA, Duesberg PH (1982) Acute leukemia virus E26 and avian myeloblastosis virus have related transformation-specific RNA sequences but different genetic structures, gene products, and oncogenic properties. Proc Natl Acad Sci USA 79: 3677–3681

Boettiger D, Dexter TM (1984a) Long-term bone marrow cultures in the analysis of viral leukemogenesis. Clin Haematol 13: 349–370

Boettiger D, Dexter TM (1984b) Src induced enhanced self-renewal of stem cells in the absence of transformation. In: Bishop JM, Rowley JD, Greaves M (eds) Genes and Cancer, UCLA Symposia on Molecular and Cellular Biology New Series, vol. 17. Liss, New York, pp 263–272

Boettiger D, Dexter TM (1986) Altered stem cell (CFU-S) function following infection of hematopoietic cells with a virus carrying v-src. Blood 67: 398–405

Boettiger D, Durban EM (1980) Progenitor cell populations can be infected by RNA tumour viruses, but transformation is dependent on the expression of specific differentiated cell functions. Cold Spring Harbor Symp Quant Biol 44: 1249–1254

Boettiger D, Durban EM (1984) Target cell for avian myeloblastosis virus in embryonic yolk sac and the relationship of cell differentiation to cell transformation. J Virol 49: 841–847

Boettiger D, Olsen M (1989) Induction of leukemia by avian myeloblastosis virus: a mechanistic hypothesis. In: Shen-Ong GLC, Potter M, Copeland NG (eds) Mechanisms in Myeloid Tumorigenesis. Curr Top Microbiol Immunol 149: 157–164

Boettiger D, Roby K, Brumbaugh J, Biehl J, Holtzer H (1977) Transformation of chicken embryo retinal melanoblasts by a temperature-sensitive mutant of Rous sarcoma virus. Cell 11: 881–890

Boettiger D, Soltesz R, Holtzer H, Pacifici M (1983) Infection of chick limb bud presumptive chondroblasts by a temperature-sensitive mutant of Rous sarcoma virus and the reversible inhibition of their terminal differentiation in culture. Mol Cell Biol 3: 1518–1526
Boettiger D, Anderson S, Dexter TM (1984) Effect of src on long-term bone marrow cultures: increased self-renewal of hematopoietic progenitor cells without leukemia. Cell 36: 763–773
Bohmann D, Bos TJ, Admon A, Nishimira T, Vogt PK, Tijan R (1987) Human proto-oncogene c-jun encodes a DNA binding protein with structural and functional properties of transcription factor AP-1. Science 232: 1386–1393
Boulter CA, Wagner EF (1988) The effects of v-src expression on the differentiation of embryonal carcinoma cells. Oncogene 2: 207–214
Bravo R, Burckhardt J, Curran T, Muller R (1985) Stimulation and inhibition of growth by EGF in different A431 cell clones is accompanied by the rapid induction of c-fos and c-myc proto-oncogenes. EMBO J 4: 1193–1197
Brelvi ZS, Studzinski GP (1987) Coordinate expression of c-myc, c-myb, and histone H4 genes in reversibly differentiating HL 60 cells. J Cell Physiol 131: 43–49
Brown AMC, Wildin RS, Pendergast TJ, Varmus HE (1986) A retrovirus vector expressing the putative mammary oncogene int-1 causes partial transformation of a mammary epithelial cell line. Cell 46: 1001–1009
Brownell E, Mathieson B, Young HA, Keller J, Ihle JN, Rice NR (1987) Detection of c-rel-related transcripts in mouse hematopoietic tissues, fractionated lymphocyte populations, and cell lines. Mol Cell Biol 7: 1304–1309
Brugge JS, Cotton PC, Queral AE, Barrett JN (1985) Neurones express high levels of a structurally modified, activated form of pp60c-src. Nature 316: 554–557
Buck CA, Horwitz AF (1987) Cell surface receptors for extracellular matrix molecules. Annu Rev Cell Biol 3: 179–206
Buick RN, Filmus J, Quaroni A (1987) Activated H-ras transforms rat intestinal epithelial cells with expression of alpha-TGF. Exp Cell Res 170: 300–309
Bunte T, Greiser Wilke I, Moelling K (1983) The transforming protein of the MC29-related virus CMII is a nuclear DNA-binding protein whereas MH2 codes for a cytoplasmic RNA-DNA binding polyprotein. EMBO J 2: 1087–1092
Bunte T, Donner P, Pfaff E, Reis B, Greiser Wilke I, Schaller H, Moelling K (1984) Inhibition of DNA binding of purified p55v-myc in vitro by antibodies against bacterially expressed myc protein and a synthetic peptide. EMBO J 3: 1919–1924
Cabrera CV, Alonso MC, Johnson P, Phillips RG, Lawrence PA (1987) Phenocopies induced with antisense RNA identify the wingless gene. Cell 50: 659–663
Calothy G, Poirier F, Dambrine G, Mignatti P, Combes P, Pessac B (1980) Expression of viral oncogenes in differentiating chick embryo neuroretinal cells infected with avian tumor viruses. Cold Spring Harbor Symp Quant Biol 44: 2-P 983–90
Carpenter G (1987) Receptors for epidermal growth factor and other polypeptide mitogens. Annu Rev Biochem 56: 881–914
Casalbore P, Agostini E, Alema S, Falcone G, Tato F (1987) The v-myc oncogene is sufficient to induce growth transformation of chick neuroretina cells. Nature 326: 188–190
Chapekar MS, Hartman KD, Knode MC, Glazer RI (1987) Synergistic effect of retinoic acid and calcium ionophore differentiation, c-myc expression, and membrane tyrosine activity in human promyelocytic leukemia cell line HL-60. Mol Pharmacol 31: 140–145
Chi JCH, Rubinstein N, Strahs K, Holtzer H (1975) Synthesis of myosin heavy and light chains in muscle cultures. J Cell Biol 67: 523–537
Clarke MF, Westin E, Schmidt D, Josephs SF, Ratner L, Wong Staal F, Gallo RC, Reitz MS Jr (1984) Transformation of NIH 3T3 cells by a human c-sis cDNA clone. Nature 308: 464–467
Clarke MF, Kukowska Latallo JF, Westin E, Smith M, Prochownik EV (1988a) Constitutive expression of a c-myb cDNA blocks Friend murine erythroleukemia cell differentiation. Mol Cell Biol 8: 884–892
Clarke MF, Kukowska-Latallo JF, Westin E, Smith M, Prochownik EV (1988b) Constitutive expression of a c-myb cDNA blocks friend murine erythroleukemia cell differentiation. Mol Cell Biol 8: 884–892

Colletta G, Cirafici AM, Vecchio G (1986) Induction of the c-fos oncogene by thyrotropic hormone in rat thyroid cells in culture. Science 233: 458–460

Conscience JF, Verrier B, Martin G (1986) Interleukin-3-dependent expression of the c-myc and c-fos proto-oncogenes in hemopoietic cell lines. EMBO J 5: 317–323

Coppola JA, Cole MD (1986) Constitutive c-myc oncogene expression blocks mouse erythroleukaemia cell differentiation but not commitment. Nature 320: 760–763

Crisanti Combes P, Lorinet AM, Girard A, Pessac B, Wasseff M, Calothy G (1982) Effects of Rous sarcoma virus on the differentiation of chick and quail neuroretina cells in culture. Adv Exp Med Biol 158: 115–122

Crisanti P, Lorinet AM, Calothy G, Pessac B (1985) Glutamic acid decarboxylase activity is stimulated in quail retina neuronal cells transformed by Rous sarcoma virus and is regulated by pp60v-src. EMBO J 4: 1467–1470

Cross FR, Garber EA, Pellman D, Hanafusa H (1984) A short sequence in the p60src N terminus is required for p60ser myristylation and membrane association and for cell transformation Mol Cell Biol 4: 1834–1842

Curran R, Teich N (1982) Candidate product of the FBJ murine ostersarcoma virus oncogene: characterization of a 55,000 dalton phosphoprotein. J Virol 42: 114–122

Curran T, Morgan JI (1985) Superinduction of c-fos by nerve growth factor in the presence. Science 229: 1265–1268

Davis RL, Weintraub H, Lassar AB (1987) Expression of a single transfected cDNA converts fibroblasts to myoblasts. Cell 51: 987–1000

Dean M, Park M, LeBeau M, Robins T, Diaz M, Rowley J, Blair D, Vande Woode GF (1985) The human met oncogene is related to the tyrosine kinase oncogenes. Nature 318: 385–388

DeClue JE, Martin GS (1987) Phosphorylation of talin at tyrosine in Rous sarcoma virus transformed cells. Mol Cell Biol 7: 371–378

Delli Bovi P, Curatola AM, Kern FG, Greco A, Ittmann M, Basilico C (1987) An oncogene isolated by transfection of Kaposi's sarcoma DNA encodes a growth factor that is a member of the FGF family. Cell 50: 729–737

Dere WH, Hiraya H, Rapoport B (1985) TSH and cAMP enhance expression of the myc proto-oncogene in cultured thyroid cells. Endocrinology 117: 2249–2251

Dexter TM (1982) Stromal cell associated haemopoiesis. J Cell Physiol [Suppl]: 87–94

Dexter TM (1984) Blood cell development. The message in the medium (News). Nature 309: 764–767

Dexter TM, Allen TD, Lajtha LG (1977a) Conditions controlling the proliferation of haemopo]etic stem cells in vitro. J Cell Physiol 91: 335–344

Dexter TM, Scott D, Teich NM (1977b) Infection of bone marrow cells in vitro with FLV: effects on stem cell proliferation, differentiation and leukamogenic capacity. Cell 12: 355–564

Dexter TM, Garland J, Scott D, et al (1980) Growth of factor-dependent hemopoietic precursor cell lines. J Exp Med 152: 1036–1047

Dickson C, Peters G (1987) Potential oncogene product related to growth factor. Nature 326: 833

Distel RJ, Ro HS, Rosen BS, Groves DL, Spiegelman BM (1987) Nucleoprotein complexes that regulate gene expression in adipocyte differentiation: direct participation of c-fos. Cell 49: 835–844

Dmitrovsky E, Kuehl WM, Hollis GF, Kirsch IR, Segal S (1986a) Expression of a transfected human c-myc oncogene inhibits differentiation of a mouse erythroleukaemia cell line. Nature 322: 748–750

Dmitrovsky E, Kuehl WM, Hollis GF, Kirsch IR, Bender TP, Segal S (1986b) A transfected c-myc oncogene inhibits mouse erythroleukemic differentiation. Curr Top Microbiol Immunol 132: 327–330

Dolberg DS, Bissell MJ (1984) Inability of Rous sarcoma virus to cause sarcomas in the avian embryo. Nature 309: 552–556

Doolittle RF, Hunkapiller MW, Hood LE, Devare SG, Robbins KC, Aaronson SA, Antoniades HN (1983) Simian sarcoma virus onc gene, v-sis, is derived from the gene or genes encoding a platelet-derived growth factor. Science 221: 275–277

Downing JR, Rettenmier CW, Sherr CJ (1988) Ligand-induced tyrosine kinase activity of the colony-stimulating factor 1 receptor in a murine macrophage cell line. Mol Cell Biol 8: 1795–1799

Duprey SP, Boettiger D (1985) Developmental regulation of c-myb in normal myeloid progenitor-cells. Proc Natl Acad Sci USA 82: 6937–6941

Durban EM, Boettiger D (1981a) Differential effects of transforming avian RNA tumor viruses on avian macrophages. Proc Natl Acad Sci US 78: 3600–3604

Durban EM, Boettiger D (1981b) Replicating differentiated macrophages can serve an target cells for transformation by avian myeloblastosis virus. J Virol 37: 488–492

Edelman GM (1986) Cell adhesion molecules in the regulation of animal form and tissue pattern. Annu Rev Cell Biol 2: 81–116

Emilia G, Donelli A, Ferrari S, Torelli U, Selleri L, Zucchini P, Moretti L, Venturelli D, Ceccherelli G (1986) Cellular levels of mRNA from c-myc, c-myb and c-fes onc-genes in normal myeloid and erythroid precursors of human bone marrow: an in sity hybridization study. Br J Haematol 62: 287–292

Endo T, Nadal Ginard B (1986) Transcriptional and posttranscriptional control of c-myc myogenesis: its mRNA remains inducible in differentiated cells and does not suppress the differentiated phenotype. Mol Cell Biol 6: 1412–1421

Endo T, Nadal Ginard B (1987) Three types of muscle-specific gene expression in fusion-blocked rat skeletal muscle cells: translational control in EGTA-treated cells. Cell 49: 515–526

Falcone G, Boettiger D, Tato F, Alema S (1984) Role of cell division in differentiation of myoblasts infected with a temperature-sensitive mutant of Rous sarcoma virus. EMBO J 3: 1327–1331

Falcone G, Tato F, Alema S (1985) Distinctive effects of the viral oncogenes myc, erb, fps, and src on the differentiation program of quail myogenic cells. Proc Natl Acad Sci USA 82: 426–430

Feldman RA, Gabrilove JL, Tam JP, Moore MA, Hanafusa H (1985) Specific expression of the human cellular fps/fes-encoded protein NCP92 in normal and leukemic myeloid cells. Proc Natl Acad Sci USA 82: 2379–2383

Ferrari S, Torelli U, Selleri L, Donelli A, Venturelli D, Moretti L, Torelli G (1985a) Expression of human c-fes onc-gene occurs at detectable levels in myeloid but not in lymphoid cell populations. Br J Haematol 59: 21–25

Ferrari S, Torelli U, Selleri L, Donelli A, Venturelli D, Narni F, Moretti L, Torelli G (1985b) Study of the levels of expression of two oncogenes, c-myc and c-myb, in acute and chronic leukemias of both lymphoid and myeloid lineage. Leuk Res 9: 833–842

Finkel MP, Biskis BO, Jinkins PB (1966) Virus induction of osteosarcomas in mice. Science 151: 698–701

Fiszman MY, Fuchs P (1975) Temperature-sensitive expression of differentiation in transformed myoblasts. Nature 254: 429–431

Foster DA, Shibuya M, Hanafusa H (1985) Activation of the transformation potential of the cellular fps gene. Cell 42: 105–115

Fraser SE, Green CR, Bode HR, Gilula NB (1987) Selective disruption of gap junctional communication interfers with patterning process in hydra. Science 237: 49–55

Friend C, Scher W, Holland JG, Sato T (1971) Hemoglobin synthesis in murine virus-induced leukemic cells in vitro: stimulation of erythroid differentiation with dimethyl sulfoxide. Proc Natl Acad Sci USA 68: 378–382

Frykberg L, Palmeri S, Beug H, Graf T, Hayman MJ, Vennstrom B (1983) Transforming capacities of avain erythroblastosis virus mutants deleted in the erbA or erbB oncogenes. Cell 32: 227–238

Fukui Y, Kozasa T, Kaziro Y, Takeda T, Yamamoto M (1986) Role of a ras homolog in the life cycle of Schizosaccharomyces pombe. Cell 44: 329–336

Fusco A, Portella G, Di Fiore PP, Berlingieri MT, Di Lauro R, Schneider AB, Vecchio G (1985) A mos oncogene-containing retrovirus, myeloproliferative sarcoma virus, transforms rat thyroid epithelial cells and irreversibly blocks their differentiation pattern. J Virol 56: 284–292

Gazit A, Igarashi H, Chiu IM, Srinivasan A, Yaniv A, Tronick SR, Robbins KC, Aaronson SA (1984) Expression of the normal human sis/PDGF-2 coding sequence induces cellular transformation. Cell 39: 89–97

Gazzolo L, Moscovici C, Moscovici MG, Samarut J (1979) Response of hematopoietic cells to avian acute leukemia viruses: effects on the differentiation of the target cells. Cell 16: 627–638

George-Weinstein M, Decker C, Horwitz A (1988) Combinations of monoclonal antibodies distinguish mesenchymal, myogenic, and chondrogenic precursors of the developing chick embryo. Dev Biol 125: 34–50

Gessler M, Barnekow A (1984) Differential expression of the cellular oncogenes c-src and c-yes in embryonal and adult chicken tissues. Biosci Rep 4: 757–770

Giotta GJ, Cohn M (1981) The expression of glial fibrillary acidic protein in a rat cerebellar cell line. J. Cell Physiol 107: 219–230

Giotta GJ, Heitzmann J, Cohn M (1980) Properties of two temperature-sensitive Rous sarcoma virus transformed cerebellar cell lines. Brain Res 202: 445–458

Giroldi L, Hollstein M, Yamasaki H (1988) Cellular-oncogene expression in Friend erythroleukemia cells: relationship to differentiation, commitment and TPA effects. Carcinogenesis 9: 817–821

Golden A, Nemeth SP, Brugge JS (1986) Blood platelets express high levels of the pp60c-src-specific tyrosine kinase activity. Proc Natl Acad Sci USA 83: 852–856

Gospodarowicz D, Ferrara N, Schweiger L, Neufeld G (1987a) Structural characterization and biological functions of Endocrine Rev 8: 95–114

Gospodarowicz D, Neufield G, Schweiger L (1987b) Fibroblast growth factor: structural and biological properties. J Cell Physiol [Suppl] 5: 15–26

Graf T, Beug H (1978) Avian leukemia viruses: interactions with their target cells in vitro and in vivo. Biochim Biophys Acta 516: 269–299

Graf T, von Kirchback A, Beug H (1981) Characterization of the hematopoetic target cells for AEV, MC29, and AMV avian leukemia viruses. Exp Cell Res 131: 331–343

Greenberger JS (1979) Phenotypically distinct target cells for murine sarcoma virus and murine leukemia virus marrow transformation in vitro. J Natl Cancer Inst 62: 337–348

Greenberger JS, Gans PJ, Davisson PB, Moloney WC (1979) In vitro induction of continuous acute promyelocytic leukemia cell lines by Friend or Abelson murine leukaemia viruses. Blood 53: 987–1001

Greenberger JS, Sakakeeny MA, Humphries RK, et al. (1983) Demonstration of permanent factor-dependent multipotential (erythroid/neutrophil/basophil) hematopoietic progenitor cell lines. Proc Natl Acad Sci USA 80: 2931–2935

Greenwald I, Sternberg PW, Horvitz HR (1983) The lin-12 locus specifies cell fates in Caenorhabditis elegans. Cell 34: 435–444

Greenwald I (1987) The lin-12 locus of Caenorhabditis elegans. BioEssays 6: 70–73

Gregory RJ, Kammermeyer KL, Vincent WS3d, Wadsworth SG (1987) Primary sequence and developmental expression of a novel Drosophila melanogaster src gene. Mol Cell Biol 7: 2119–2127

Griffin CA, Baylin SB (1985) Expression of the c-myb oncogene in human small cell lung carcinoma. Cancer Res 45: 272–275

Gross JL, Rifkin DFB (1979) The effect of avian retroviruses on limb bud chondrogenesis in vitro. Cell 18: 707–718

Gross L (1970) Oncogenic viruses 2nd eds. Pergamon, New York

Hafen E, Basler K, Edstrom JE, Rubin GM (1987) sevenless, a cell specific homeotic gene of Drosophila, encodes a putative transmembrane receptor with a tyrosine kinase domain. Science 236: 55–63

Hall DE, Neugebauer M, Reichardt LF (1987) Embryonic neural retinal cell response to extracellular matrix proteins: developmental changes and effects of the cell substratum attachment antibody (CSAT). J Cell Biol 104: 623–634

Hamaguchi M, Hanafusa H (1987) Association of p60src with Triton X-100-resistant cellular structure correlates with morphological transformation. Proc Natl Acad Sci USA 84: 2312–2316

Hankins WD, Scolnick EM (1981) Harvey and Kirsten sarcoma viruses promote the growth and differentiation of erythroid precursor cells in vitro. Cell 26: 91–97

Hapel AJ, Vande Woode GF, Campbell HD, Young IG, Robins T (1987) Generation of an autocrine leukemia using a retroviral expression vector carrying the interleukin-3 gene. Lymphokine Res 5: 249–254

Hara K, Suda T, Suda J, Eguchi M, Ihle JN, Nagata S, Miura Y, Saito M (1988) Bipotential murine hemopoietic cell line (NFS-60) that is responsive to IL-3, GM-CSF, and erythropoietin. Exp Hematol 16: 256–261

Harris AW, Pinkert CA, Crawford M, Langdon WY, Brinster RL, Adams JM (1988) The E mu-myc transgenic mouse. A model for high-incidence spontaneous lymphoma and leukemia of early B cells. J Exp Med 167: 353–371

Heard JM, Roussel MF, Rettenmier CW, Sherr CJ (1987a) Multilineage haematopoietic disorders induced by transplantation of bone marrow cell expressing the v-fms oncogene. Cell 51: 663–673

Heard JM, Roussel MF, Rettenmier CW, Sherr CJ (1987b) Synthesis, post-translational processing, and autocrine transforming activity of a carboxylterminal truncated form of colony stimulating factor-1. Oncogene Res 1: 423–440

Henkemeyer MJ, Gertler FB, Goodman W, Hoffmann FM (1987) The drosophila abelson proto-oncogene homolog: identification of mutant alleles that have pleiotropic effects late in development. Cell 51: 821–828

Henkemeyer MJ, Bennett RL, Gertler FB, Hoffmann FM (1988) DNA sequence, structure, and tyrosine kinase activity of the Drosophila melanogaster Abelson proto-oncogene homolog. Mol Cell Biol 8: 843–853

Hill CS, Duran S, Lin Z, Weber K, Holtzer H (1986) Titin and myosin, but not desmin, are linked during myofibrillogenesis in postmitotic mononucleated myoblasts. J Cell Biol 103: 2185–2196

Hirst R, Horwitz A, Buck C, Rohrschneider L (1986) Phosphorylation of the fibronectin receptor complex in cells transformed by oncogenes that encode tyrosine kinases. Proc Natl Acad Sci USA 83: 6470–6474

Holmes KL, Pierce JH, Davidson WF, Morse HC3d (1986) Murine hematopoietic cells with pre-B or pre-B/myeloid characteristics are generated by in vitro transformation with retroviruses containing fes, ras, abl, and src oncogenes. J Exp Med 164: 443–457

Holt JT, Redner RL, Nienhuis AW (1988) An oligomer complementary to c-myc mRNA inhibits proliferation of HL-60 promyelocytic cells and induces differentiation. Mol Cell Biol 8: 963–973

Holtzer H, Beihl J, Yeoh G, Meganathan R, Kaji A (1975) Effects of oncogenic virus on muscle differentiation. Proc Natl Acad Sci USA 72: 4051–4055

Holtzer H, Biehl J, Antin P, Tokunaka S, Sasse J, Pacifici M, Holtzer S (1983) Quantal and proliferative cell cycles: how lineages generate cell diversity and maintain fidelity. Prog Clin Biol Res 134: 213–227

Huebner K, ar Rushdi A, Griffin CA, Isobe M, Kozak C, Emanuel BS, Nagarajan L, Cleveland JL, Bonner TI, Goldsborough MD, et al. (1986) Actively transcribed genes in the raf oncogene group, located on the X chromosome in mouse and human. Proc Natl Acad Sci USA 83: 3934–3938

Hynes RO (1987) Integrins: a family of cell surface receptors. Cell 48: 549–554

Iba H, Jove R, Hanafusa H (1985) Lack of induction of neuroretinal cell proliferation by Rous sarcoma virus variants that carry the c-src gene. Mol Cell Biol 5: 2856–2859

Ihle JN (1986) Immunological regulation of hematopoietic/lymphoid stem cell differentiation by interleukin 3. Adv Immunol 39: 1–50

Ikebuchi K, Clark SC, Ihle JN, Souza LM, Ogawa M (1988) Granulocyte colony-stimulating factor enhances interleukin 3-dependent proliferation of multipotential hemopoietic progenitors. Proc Natl Acad Sci USA 85: 3445–3449

Introna M, Hamilton TA, Kaufman RE, Adams DO (1986) Treatment of murine peritoneal macrophages with bacterial lipopolysaccharide alters expression of c-fos and c-myc oncogenes. J Immunol 137: 2711–2715

Jaffe BD, Sabath DE, Johnson GD, Moscinski LC, Johnson KR, Rovera G, Nauseef WM, Prystowsky MB (1988) Myeloperoxidase and oncogene expression in GM-CSF induced bone marrow differentiation. Oncogene 2: 167–174

Johnson GR (1984) Haemopoietic multipotential stem cells in culture. Clin Haematol 13: 309–327

Jove R, Garber EA, Iba H, Hanafusa H (1986) Biochemical properties of p60v-src mutants that induce different cell transformation parameters. J Virol 60: 849–857

Jurdic P, Moscovici C, Pessano S, Bottero L, Rovera G (1984) A monoclonal antibody with specificity for leukemic cells transformed by defected avian leukemia viruses. J Cell Physiol 2: 85–95

Kahn P, Frykberg L, Brady C, Stanley I, Beug H, Vennstrom B, Graf T (1986) v-erbA cooperates with sarcoma oncogenes in leukemic cell transformation. Cell 45: 349–356

Kamps MP, Sefton BM (1988) Identification of multiple novel polypeptide substrates of the v-src, v-yes, v-fps, v-ros, and v-erbB oncogenic tyrosine kinases utilizing antisera against phosphotyrosine. Oncogene 2: 305–315

Kamps MP, Buss JE, Sefton BM (1985) Mutation of NH2-terminal glycine of p60src prevents both myristoylation and morphological transformation. Proc Natl Acad Sci USA 82: 4625–4628

Kan NC, Baluda MA, Papas TS (1985) Sites of recombination between the transforming gene of avian myeloblastosis virus and its helper virus. Virology 145: 323–329

Kanter MR, Smith RE, Hayward WS (1988) Rapid induction of B-cell lymphoma: insertional activation of c-myb by avian leukosis virus. J Virol 62: 1423–1432

Kardami E, Spector D, Strohman RC (1988) Heparin inhibits skeletal muscle growth in vitro. Dev Biol 126: 19–28

Katzen AL, Kornberg TB, Bishop JM (1985) Isolation of the proto-loncogene c-myb from D. melanogaster. Cell 41: 449–456

Kaufmann Y, Silverman T, Levi BZ, Ozato K (1987) Induction of c-ets and c-fos gene expression upon antigenic stimulation of a T cell hybridoma with inducible cytolytic capacity. J Exp Med 166: 810–815

Kimelman D, Kirschner M (1987) Synergistic induction of mesoderm by FGF and TGF-B and the identification of an mRNA coding for FGF in the early Xenopus embryo. Cell 51: 869–877

King MW, Roberts JM, Eisenman RN (1986) Expression of the c-myc proto-oncogene during development of xenopus laevis. Mol Cell Biol 6: 4499–4508

Kirsch IR, Bertness V, Silver J, Hollis GF (1986) Regulated expression of the c-myb and c-myc oncogenes during erythroid differentiation. J Cell Biochem 32: 11–21

Kitamura N, Kitamura A, Toyoshima K, Hirayama Y, Yoshida M (1982) Avian sarcoma virus Y73 genome sequence and structural similarity of its transforming gene product to that of Rous sarcoma virus. Nature 297: 205–207

Kleiman HK, Kleve RF, Martin GR (1981) Role of collagenous matrices in the adhesion and growth of cells. J Cell Biol 88: 473–485

Klempnauer KH, Symonds G, Evan GI, Bishop JM (1984) Subcellular localization of proteins encoded by oncogenes of avian myeloblastosis virus and avian leukemia virus E26 and by chicken c-myb gene. Cell 37: 537–547

Knight KL, Spieker Polet H, Kazdin DS, Oi VT (1988) Transgenic rabbits with lymphocytic leukemia induced by the c-myc oncogene fused with the immunoglobulin heavy chain enhancer. Proc Natl Acad Sci USA 85: 3130–3134

Kovacs EJ, Oppenheim JJ, Young HA (1986) Induction of c-fos and c-myc expression in T lymphocytes after treatment with recombinant interleukin 1-alpha. J Immunol 137: 3649–3651

Kruijer W, Cooper JA, Hunter T, Verma IM (1984) Platelet-derived growth factor induces rapid but transient expression of the c-fos gene and protein. Nature 312: 711–716

Kruijer W, Schubert D, Verma IM (1985) Induction of the proto-oncogene fos by nerve growth factor. Proc Natl Acad Sci USA 82: 7330–7334

Lacy J, Sarkar SN, Summers WC (1986) Induction of c-myc expression in human B lymphocytes by B-cell growth factor and anti-immunoglobulin. Proc Natl Acad Sci USA 83: 1458–1462

Land H, Chen AC, Morgenstern JP, Parada LF, Weinberg RA (1986) Behavior of myc and ras oncogenes in transformation of rat embryo fibroblasts. Mol Cell Biol 6: 1917–1925

Langdon WY, Harris AW, Cory S, Adams JM (1986) The c-myc oncogene perturbs B lymphocyte development in E-mu-myc transgenic mice. Cell 47: 11–18

Langer-Safer PR, Lehrman SR, Skalka AM (1985) v-src inhibits differentiating via an extracellular intermediate(s). Mol Cell Biol 5: 2847–2850

Lax I, Kris R, Sasson I, Ullrich A, Hayman MJ, Beug H, Schlessinger J (1985) Activation of c-erbB in avian leukosis virus-induced erythroblastosis leads to the expression of a truncated EGF receptor kinase. EMBO J 4: 3179–3182

Leder A, Pattengale PK, Kuo A, Stewart TA, Leder P (1986) Consequences of widespread deregulation of the c-myc gene in transgenic mice: multiple neoplasms and normal development. Cell 45: 485–495

Lehmann R, Jimenez F, Dietrich V, Campos-Ortega JA (1983) On the phenotype and development of mutants of early neurogenesis in Drosophila melanogaster. Wilhelm Roux's Arch Dev Biol 192: 62–74

Leon J, Guerrero I, Pellicer A (1987) Differential expression of the ras gene family in mice. Mol Cell Biol 7: 1535–1540

Lev Z, Leibovitz N, Segev O, Shilo BZ (1984) Expression of the src and abl cellular oncogenes during development of Drosophila melanogaster. Mol Cell Biol 4: 982–984

Lev Z, Kimchie Z, Hessel R, Segev O (1985) Expression of ras cellular oncogenes during development of Drosophila melanogaster. Mol Cell Biol 5: 1540–1542

Levy JB, Dorai T, Wang L, Brugge JS (1987) The structurally distinct form of pp60 c-src detected in neuronal cells is encoded by a unique c-src mRNA. Mol Cell Biol 7: 4142–4145

Li Y, Turck CM, Teumer JK, Stavnezer E (1986) Unique sequence, ski, in Sloan-Kettering avian retroviruses with properties of a new cell-derived oncogene. J Virol 57: 1065–1072

Linder ME, Burr JG (1988) Nonmyristoylated p60v-src fails to phosphorylate proteins of 115–120 kDa in chicken embryo fibroblasts. Proc Natl Acad Sci USA 85: 2608–2612

Lipsich L, Brugge JS, Boettiger D (1984) Expression of the Rous sarcoma virus src gene in avian macrophages fails to elicit transformed cell phenotype. Mol Cell Biol 4: 1420–1424

Lowy DR, Willumsen BM (1986) The ras gene family. Cancer Surv 5: 275–289

Macara IG (1985) Oncogenes, ions, and phospholipids. Am J Physiol 248: C3–C11

MacDonald I, Levy J, Pawson T (1985) Expression of the mammalian c-fes protein in hematopoietic cells and identification of a distinct fes-related protein. Mol Cell Biol 5: 2543–2551

Mahon KA, Chepelinsky AB, Khillan JS, Overbeek PA, Piatigorsky J, Westphal H (1987) Oncogenesis of the lens in transgenic mice. Science 235: 1622–1628

Maki Y, Bos TJ, Davis C, Starbuck M, Vogt PK (1987) Avian sarcoma virus 17 carries the jun oncogene. Proc Natl Acad Sci USA 84: 2848–2852

Maness PF, Sorge LK, Fults DW (1986a) An early developmental phase of pp60c-src expression in the neural ectoderm. Dev Biol 117: 83–89

Maness PF (1986b) pp60c-src encoded by the proto-oncogene c-src is a product of sensory neurons. J Neurosci Res 16: 127–139

Mark GE, Seeley TW, Shows TB, Mountz JD (1986) Pks, a raf-related sequence in humans. Proc Natl Acad Sci USA 83: 6312–6316

Mark GE, MacIntyre RJ, Digan ME, Ambrosio L, Perrimon N (1987) Drosophila melanogaster homologs of the raf oncogene. Mol Cell Biol 7: 2134–2140

Martinet Y, Bitterman PB, Mornex JF, Gortendorst GR, Martin GR, Crystal RG (1986) Activated human monocytes express the c-sis proto-oncogene and release a mediator showing PDGF-like activity. Nature 319: 158–160

Massague J (1987) The TGF-beta family of growth and differentiation factors. Cell 49: 437–438

McDonough SK, Larsen S, Brodey RS, Stock ND, Hardy WDJr (1971) A transmissible feline fibrosarcoma of viral origin. Cancer Res 31: 953–956

Mellersh H, Strain AJ, Hill DJ (1986) Expression of the proto-oncogenes C-H-ras and N-ras in early trimester human fetal tissues. Biochem Biophys Res Commun 141: 510–516

Menko AS, Boettiger D (1987) Occupation of the extracellular matrix receptor integrin is a control point for myogenic differentiation. Cell 51: 51–57

Menko AS, Boettiger D (1988) Inhibition of chick embryo lens differentiation and lens junction formation in culture by pp60v-src. Mol Cell Biol 8: 1414–1420

Menko AS, Toyama Y, Boettiger D, Holtzer H (1983) Altered cell spreading in cytochalesin B: a possible role for intermediate filaments. Mol Cell Biol 3: 113–125

Menko AS, Klukas K, Johnson R (1984) Chicken embryo lens cultures mimic differentiation in the lens. Dev Biol 103: 129–141

Metcalf D (1977) Hemopoietic colonies. In vitro cloning of normal and leukemic cells. Springer, Berlin Heidelberg New York

Metcalf D (1986) Haemopoietic growth factors now cloned. Br J Haematol 62: 409–412

Milbrandt J (1986) Nerve growth factor rapidly induces c-fos mRNA in PC12 rat pheochromocytoma cells. Proc Natl Acad Sci USA 83: 4789–4793

Moelling K, Heimann B, Beimling P, Rapp UR, Sander T (1984) Serine- and threonine-specific protein kinase activities of purified gag-mil and gag-raf proteins. Nature 312: 558–561

Moelling K, Pfaff E, Beug H, Beimling P, Bunte T, Graf T (1985) DNA-binding activity is associated with purified myb proteins from AMV and E26 viruses and is temperature-sensitive for E26 ts mutants. Cell 40: 983–990

Moscovici C (1967) A quantitative assay for avian myeloblastosis virus. Proc Soc Exp Biol Med 125: 1213–1215

Moscovici MG, Jurdic P, Samarut J, Gazzolo L, Mura CV, Moscovici C (1983) Characterization of the hematopoietic target cells for the avian leukemia virus E26. Virology 129: 65–78

Moss P, Radke K, Carter VC, Young J, Gilmore T, Martin GS (1984) Cellular localization of the transforming protein of wild-type and temperature-sensitive Fujinami sarcoma virus. J Virol 52: 557–565

Muller R, Verma IM (1984) Expression of cellular oncogenes. Curr Top Microbiol Immunol 112: 73–115

Nadal Ginard B (1978) Committment, fusion and biochemical differentiation of a myogenic cell line in the absence of DNA synthesis. Cell 15: 855–864

Naharro G, Tronick SR, Rasheed S, Gardner MB, Aaronson SA, Robbins KC (1983) Molecular cloning of integrated Gardner-Rasheed feline sarcoma virus: genetic structure of its cell-derived sequence differs from that of other tyrosine kinase-coding onc genes. J Virol 47: 611–619

Nakagawa T, Mabry M, de Bustros A, Ihle JN, Nelkin BD, Baylin SB (1987) Introduction of v-Ha-ras oncogene induces differentiation of cultured human medullary thyroid carcinoma cells. Proc Natl Acad Sci USA 84: 5923–5927

Neckameyer WS, Shibuya M, Hsu MT, Wang LH (1986) Proto-oncogene c-ros codes for a molecule with structural common to those of growth factor receptors and displays tissue specific and developmentally regulated expression. Mol Cell Biol 6: 1478–1486

Ness SA, Beug H, Graf T (1987) v-myb dominance over v-myc in doubly transformed chick myelomonocytic cells. Cell 51: 41–50

Nilsson K, Larsson LG, Carlsson M, Danersund A, Hellman L, Totterman T, Pettersson U (1986) Expression of c-myc and c-fos during phorbol ester induced differentiation of B-type chronic lymphocytic leukemia cells. Curr Top Microbiol Immunol 132: 280–289

Nishizawa M, Semba K, Yoshida MC, Yamamoto T, Sasaki M, Toyoshima K (1986) Structure, expression, and chromosomal location of the human c-fgr gene. Mol Cell Biol 6: 511–517

Notter MF, Navon SE, Fung BK, Balduzzi PC (1987) Infection of neuroretinal cells in vitro by avian sarcoma viruses UR1 and UR2: transformation, cell growth stimulation, and changes in transducin levels. Virology 160: 489–493

Nunn M, Weiher H, Bullock P, Duesberg P (1984) Avian erythroblastosis virus E26: nucleotide sequence of the tripartite onc gene and of the LTR, and analysis of the cellular prototype of the viral ets sequence. Virology 139: 330–339

Ogawa M, Porter PN, Nakahata T (1983) Renewal and committment to differentiation of hemopoietic stem cells (an interpretative review). Blood 61: 823–829

Olson EN, Spizz G, Tainsky MA (1987) The oncogenic forms of N-ras or H-ras prevent skeletal myoblast differentiation. Mol Cell Biol 7: 2104–2111

Ong J, Yamashita S, Melmed S (1987) Insulin-like growth factor I induces c-fos messenger ribonucleic acid in L6 rat skeletal muscle cells. Endocrinology 120: 353–357

Overell RW, Watson JD, Gallis B, Weisser KE, Cosman D, Widmer MB (1987) Nature and specificity of lymphokine independence induced by a selectable retroviral vector expressing v-src. Mol Cell Biol 7: 3394–3401

Pacifici M, Boettiger D, Roby K, Holtzer H (1977) Transformation of chondroblasts by Rous sarcoma virus and the synthesis of the extracellular matrix. Cell 11: 891–899

Papkoff J, Nigg EA, Hunter T (1983) The transforming protein of Moloney murine sarcoma virus is a soluble cytoplasmic protein. Cell 33: 161–172

Pasquale EB, Maher PA, Singer SJ (1986) Talin is phosphorylated on tyrosine in chicken embryo fibroblasts transformed by Rous sarcoma virus. Proc Natl Acad Sci USA 83: 5507–5511

Payne PA, Olson EN, Hsiau P, Roberts R, Perryman MB, Schneider MD (1987) An activated c-Ha-ras allele blocks the induction of muscle-specific genes whose expression is contingent on mitogen withdrawal. Proc Natl Acad Sci USA 84: 8956–8960

Pfeifer Ohlsson S, Rydnert J, Goustin AS, Larsson E, Betsholtz C, Ohlsson R (1985) Cell-type-specific pattern of myc proto oncogene expression in developing human embryos. Proc Natl Acad Sci USA 82: 5050–5054

Pierce JH, DiFiore PP, Aaronson SA, Potter M, Pumphrey A, Scott A, Ihle JN (1985) Neoplastic transformation of mast cells by Abelson-MuLV: abrogation of IL-3 dependence by a non-autocrine mechanism. Cell 41: 685–693

Pribyl LJ, Watson DK, McWilliams MJ, Ascione R, Papas TS (1988) The Drosophila ets-2 gene: molecular structure, chromosomal localization, and developmental expression. Dev Biol 127: 45–53

Prywes R, Foulkes JG, Rosenberg N, Baltimore D (1983) Sequences of the A-MuLV protein needed for fibroblast and lymphoid cell transformation. Cell 34: 569–579

Quaife CJ, Pinkert CA, Ornitz DM, Palmiter RD, Brinster RL (1987) Pancreatic neoplasia induced by ras expression in acinar cells of transgenic mice. Cell 48: 1023–1034

Radke K, Beug H, Kornfeld S, Graf T (1982) Transformation of both erythroid and myeloid cells by E26, an avian leukemia virus that contains the myb gene. Cell 31: 643–53

Ran W, Dean M, Levine RA, Henkle C, Campisi J (1986) Induction of c-fos and c-myc mRNA by epidermal growth factor or calcium ionophore is cAMP dependent. Proc Natl Acad Sci USA 83: 8216–8220

Raynaud I, Biquard JM, Chambard P, Fasciotto B, Samarut J, Blanchet JP, Krsmanovic V (1987) AEV-transformed erythroleukemia cell induced differentiation: expression of specific cell membrane antigenic molecules. Arch Virol 93: 213–222

Reed JC, Sabath DE, Hoover RG, Prystowsky MB (1985) Recombinant interleukin 2 regulates levels of c-myc mRNA in a murine T lymphocyte. Mol Cell Biol 5: 3361–3368

Reed JC, Alpers JD, Scherle PA, Hoover RG, Nowell PC, Prystowsky MB (1987) Proto-oncogene expression in cloned T lymphocytes: mitogens and growth factors induce different patterns of expression. Oncogene 1: 223–228

Rein A, Keller J, Schultz AM, Holmes KL, Medicus R, Ihle JN (1985) Infection of immune mast cells by Harvey sarcoma virus: immortalization without loss of requirement for interleukin-3. Mol Cell Biol 5: 2257–2264

Reynolds RK, Hoekzema GS, Vogel J, Hinrichs SH, Jay G (1988) Multiple endocrine neoplasia induced by the promiscuous expression of a viral oncogene. Proc Natl Acad Sci USA 85: 3135–3139

Rhim JS, Jay G, Arnstein P, Price FM, Sanford KK, Aaronson SA (1985) Neoplastic transformation of human epidermal keratinocytes by AD12-SV40 and Kirsten sarcoma viruses. Science 227: 1250–1252

Rijsewijk F, Scheermann M, Wagenaar E, Paren P, Weigel D, Nusse R (1987) The drosopholia homolog of the mouse mammary oncogene int-1 is identical to the segment polarity gene wingless. Cell 50: 647–657

Rohrschneider L, Rosok MJ, Gentry LE (1983) Molecular interaction of the src gene product with cellular adhesion plaques. Prog Nucleic Acid Res Mol Biol 29: 233–244

Rong PM, Ziller C, Penna-Melian A, Le Douarin N (1987) A monoclonal antibody specific for early avian myogenic cells and differentiated muscle. Dev Biol 122: 338–353

Rosenberg N, Baltimore D (1976) A quantitative assay for transformation of bone marrow cells by abelson murine leukemia virus. J Exp Med 143: 1453–1463

Roussel M, Dull TJ, Rettenmier CW, Ralph P, Ullrich A, Sherr CJ (1987) Transforming potential of the c-fms proto-oncogene (CFS-1 receptor). Nature 325: 549–552

Roy Burman P, Devi BG, Parker JW (1983) Differential expression of c-erbB, c-myc and c-myb oncogene loci in human lymphomas and leukemias. Int J Cancer 32: 185–191

Ruther U, Garber C, Komitowski D, Muller .R, Wagner EF (1987) Deregulated c-fos expression interferes with normal bone development in transgenic mice. Nature 325: 412–416

Sacca R, Stanley ER, Sherr CJ, Rettenmier CW (1986) Specific binding of the mononuclear phagocyte colony-stimulating factor CSF-1 to the product of the v-fms oncogene. Proc Natl Acad Sci USA 83: 3331–3335

Samarut J, Gazzolo L (1982) Target cells infected by erythroblastosis virus differentiats and become transformed. Cell 28: 921–929

Sap J, Munoz A, Damm K, Goldberg Y, Ghysdael J, Leutz A, Beug H, Vennstrom B (1986) The c-erb-A protein is a high-affinity receptor for thyroid hormone. Nature 324: 635–640

Sariban E, Mitchell T, Kufe D (1985) Expression of the c-fms proto-oncogene during human monocytic differentiation. Nature 316: 64–66

Schartl M, Barnekow A (1984) Differential expression of the cellular src gene during vertebrate development. Dev Biol 105: 415–422

Schmidt A, Setoyama C, de Crombrugghe B (1985) Regulation of a collagen gene promoter by the product of viralmos oncogene. Nature 314: 286–289

Schoenenberger CA, Andres AC, Groner B, van der Valk M, LeMeur M, Gerlinger P (1988) Targeted c-myc gene expression in mammary glands of transgenic mice induces mammary tumours with constitutive milk protein gene transcription. EMBO J 7: 169–175

Schwartz RI, Farson DA, Soo W, Bissel MJ (1978) Primary avian tendon cells in culture: an improved system for understanding malignant transformation. J Cell Biol 79: 672–679

Sefton BM, Hunter T, Ball EH, Singer SJ (1981) Vinculin: a cytoskeletal target for the transforming protein of Rous sarcoma virus. Cell 24: 165–174

Segal D, Shilo BZ (1986a) Tissue localization of Drosophila melanogaster ras transcripts during development. Mol Cell Biol 6: 2241–2248

Segal D, Shilo BZ (1986b) Tissue localization of Drosophila melanogaster ras transcripts during development. Mol Cell Biol 6: 2241–2248

Semba K, Nishizawa M, Miyajima N, Yoshida MC, Sukegawa J, Yamanashi Y, Sasaki M, Yamamoto T, Toyoshima K (1986) Yes-related protooncogene, syn, belongs to the protein-tyrosine kinase family. Proc Natl Acad Sci USA 83: 5459–5463

Seth A, Priel E, Vande Woude GF (1987) Nucleoside triphosphate-dependent DNA-binding properties of mos protein. Proc Natl Acad Sci USA 84: 3560–3564

Setoyama C, Liau G, de Crombrugghe B (1985) Pleiotropic mutants of NIH 3T3 cells with altered regulation in the expression of both type I collagen and fibronectin. Cell 41: 201–209

Shackleford GM, Varmus HE (1987) Expression of the proto-oncogene int-1 is restricted to postmeiotic male germ cells and the neural tube of mid-gestational embryos. Cell 50: 89–95

Shanley DJ, Cossu G, Boettiger D, Holtzer H, Pacifici M (1983) Transformation by Rous sarcoma virus induces similar patterns of glycosaminoglycan synthesis in chick embryo skin fibroblasts and vertebral chondroblasts. J Biol Chem 258: 810–816

Sheiness D, Gardinier M (1984a) Expression of a proto-oncogene (proto-myb) in hemopoietic tissues of mice. Mol Cell Biol 4: 1206–1212

Sheiness D, Gardinier M (1984b) Expression of a proto-oncogene (proto-myb) in hemopoietic tissues of mice. Mol Cell Biol 4: 1206–1212

Shen-Ong GLC, Holmes KL, Morse HC III (1987) Phorbol ester-induced growth arrest of murine myelomonocytic leukemic cells with virus-disrupted myb locus is not accompanied by decreased myc and myb expression. Proc Natl Acad Sci USA 84: 199–203

Sherr CJ, Rettenmier CW, Sacca R, Roussel MF, Look AT, Stanley ER (1985) The c-fms proto-oncogene product is related to the receptor for the mononuclear phagocyte growth factor receptor, CSF-1, Cell 41: 665–676

Sherr CJ (1987) Fibroblast and hematopoietic cell transformation by the fms oncogene (CSF-1 receptor). J Cell Physiol [Suppl] 5: 83–88

Shipp MA, Reinherz EL (1987) Differential expression of nuclear proto-oncogenes in T cells triggered with mitogenic and nonmitogenic T3 and T11 activation signals. J Immunol 139: 2143–2148

Simon MA, Drees B, Kornberg T, Bishop JM (1985) The nucleotide sequence and the tissue-specific expression of Drosophila c-src. Cell 42: 831–840

Skouv J, Christensen B, Skibshj I, Autrup H (1986) The skin tumor-promoter 12-O-tetradecanoylphorbol-13-acetate induces transcription of the c-fos proto-oncogene in human bladder epithelial cells. Carcinogenesis 7: 331–333

Slamon DJ, Boone TC, Murdock DC, Keith DE, Press MF, Larson RA, Souza LM (1986) Studies of the human c-myb gene and its product in human acute leukemias. Science 233: 347–351

Smith MR, DeGudicibus SJ, Stacey DW (1986) Requirement for c-ras proteine during viral oncogene transformation. Nature 320: 540–543

Sorge LK, Levy BT, Maness PF (1984) pp60c-src is developmentally regulated in the neural retina. Cell 36: 249–257

Sorge JP, Sorge LK, Maness PF (1985) pp60c-src is expressed in human fetal and adult brain. Am J Pathol 119: 151–157

Spiz G, Hu J, Olson EN (1987) Inhibition of myogenic differentiation by fibroblast growth factor or type B differentiation by fibroblast growth factor or type B transforming growth factor does not require persistent c-myc expression. Dev Biol 123: 500–507

Spooner E, Boettiger D, Dexter TM (1984) Continuous in vitro generation of multipotential stem cell clones fron src infected cultures. Nature 310: 228–230

Spooner E, Heyworth CM, Dunn A, Dexter TM (1986) Self-renewal and differentiation of interleukin-3-dependent multipotent stem cells are modulated by stromal cells and serum factors. Differentiation 31: 111–118

Staron RS, Patte D (1987) The multiplicity of combinations of myosin light chains and heavy chains in histochemically typed single fibres. Rabbit tibialis anterior muscle. Biochem J 243: 695–699

Stewart CL, Schuetze S, Vanek M, Wagner EF (1987) Expression of retroviral vectors in transgenic mice obtained by embryo infection. EMBO J 6: 383–388

Studzinski GP, Bhandal AK, Brelvi ZS (1985) A system for monocytic differentiation of leukemic cells HL 60 by a short exposure to 1,25-dihydroxycholecalciferol. Proc Soc Exp Biol Med 179: 288–295

Studzinski GP, Brelvi ZS, Feldman SC, Watt RA (1986) Participation of c-myc protein in DNA synthesis of human cells. Science 234: 467–470

Suda Y, Aizawa S, Hirai S, Inoue T, Furuta Y, Suzuki M, Hirohashi S, Ikawa Y (1987) Driven by the same Ig enhancer and SV40 T promoter ras induced lung adenomatous tumors, myc induced pre-B cell lymphomas and SV40 large T gene a variety of tumors in transgenic mice. EMBO J 6: 4155–4065

Sullivan NF, Watt RA, Delannoy MR, Green CL, Spector DL (1986) Colocalization of the myc oncogene protein and small nuclear ribonucleoprotein particles. Cold Spring Harbor Symp Quant Biol 51: 943–947

Taira M, Yoshida T, Miyagawa K, Sakamoto H, Terada M, Sugimura T (1987) cDNA sequence of human transforming gene hst and identification of the coding sequence required for transforming activity. Proc Natl Acad Sci USA 84: 2980–2984

Teich NM, Dexter TM (1978) Effects of murine leukemia virus infection on the differentiation of hemapoietic cells in culture. In: Clarkson B, Marks PA, Till JE (eds) Differentiation of normal and neoplastic hematopoietic cells. Cold Spring Harbor Laboratory, Cold Spring Harbor pp 657–670

Till JE, McCulloch EA (1961) A direct measurement of the radiation sensitivity of normal mouse bone marrow cells. Radiat Res 14: 213–222

Tushinski RJ, Oliver IT, Guilbert LJ, Stanley ER (1982) Survival of mononuclear phagocytes depends on a lineage-specific growth factor that the differentiated cells selectively destroy. Cell 28: 81–91

Umemura T, Umene K, Takahira H, Takeichi N, Katsuno M, Fukumaki Y, Nishimura J, Sakaki Y, Ibayashi H (1988) Hematopoietic growth factors (BPA and Epo) induce the expressions of c-myc and c-fos proto-oncogenes in normal human erythroid progenitors. Leuk Res 187–194

Uzvolgyi E, Kiss I, Pitt A, Arsenian S, Ingvarsson S, Udvardy A, Hamada M, Klein G, Sumegi J (1988) Drosophila homolog of the murine Int-1 protooncogene. Proc Natl Acad Sci USA 85: 3034–3038

Velu TJ, Beguinot L, Vass WC, Willingham MC, Merlino GT, Pastan I, Lowy DR (1987) Epidermal growth factor-dependent transformation by a human EGF receptor proto-oncogene. Science 235: 1408–1410

Vogt PK, Bos TJ, Doolittle RF (1987) Homology between the DNA-binding domain of the GCN4 regulatory protein of yeast and the carboxyl-terminal region of a protein coded for by the oncogene jun. Proc Natl Acad Sci USA 84: 3316–3319

Vogt TF, Solter D, Tilghman SM (1987) Raf, a trans-acting locus, regulates the alpha-fetoprotein gene in a cell-autonomous manner. Science 236: 301–303

Voronova AF, Sefton BM (1986) Expression of a new tyrosine protein kinase is stimulated by retrovirus promoter insertion. Nature 319: 682–685

Wadsworth SC, Madhavan K, Bilodeau Wentworth D (1985) Maternal inheritance of transcripts from three Drosophila src-related genes. Nucleic Acids Res 13: 2153–2170

Waneck GL, Rosenberg N (1981) Abelson leukemia virus induces lymphoid and erythroid colonies in infected fetal liver cultures. Cell 26: 91–97

Warner AE, Guthrie SC, Gilula NB (1984) Antibodies to gap junctional protein selectively disrupt junctional communication. Nature 311: 127–129

Watson DK, McWilliams-Smith MJ, Nunn MF, Duesberg PH, O'Brien SJ, Papas TS (1985) The ets sequence from the transforming gene of avian erythroblastosis virus, E26, has unique domains on human chromosomes 11 and 21: both loci are transcriptionally active. Proc Natl Acad Sci USA 82: 7294–7298

Watson JD, Eszes M, Overell R, Conlon P, Widmer M, Gillis S (1987) Effect of infection with murine recombinant retroviruses containing the v-src oncogene on interleukin 2- and interleukin 3-dependent growth states. J Immunol 139: 123–129

Watt RA, Shatzman AR, Rosenberg M (1985) Expression and characterization of the human c-myc DNA binding protein. Mol Cell Biol 5: 448–456

Weeks DL, Melton DA (1987) A maternal mRNA localized to the vegelal hemisphere in Xenopus eggs codes for a growth factor related to TGF-B. Cell 51: 861–867

Weinberger C, Giguere V, Hollenberg SM, Thompson C, Arriza J, Evans RM (1987) Human steroid receptors and erb-A gene products form a superfamily of enhancer-binding proteins. Clin Physiol Biochem 5: 179–189

Weiss RA, Teic N, Varmus H, Coffin J (1982) RNA Tumor Viruses, 2nd edn. Cold Spring Harbor Laboratory, Cold Spring Harbor

Weissman B, Aaronson SA (1985) Members of the src and ras oncogene families supplant the epidermal growth factor requirement of BALB/MK-2 keratinocytes and induce distinct alterations in their terminal differentiation program. Mol Cell Biol 5: 3386–3396

West CM, Boettiger D (1982) Selective effect of Rous sarcoma virus src gene expression on contractille protein systhesis in chick embryo myotubes. Cancer Res. 43: 2042–2048

Wharton KA, Johansen KM, Xu T, Artavanis-Taskonas S (1985) Nucleotide sequence from the neurogenic locus notch implies a gene product which shares homology with proteins containing EGF-like repeats. Cell 43: 567–581

Wheeler EF, Rettenmier CW, Look AT, Sherr CJ (1986) The v-fms oncogene induces factor independence and tumorigenicity in CSF-1 dependent macrophage cell line. Nature 324: 377–380

Wheeler EF, Askew D, May S, Ihle JN, Sherr CJ (1987) The v-fms oncogene induces factor independent growth and transformation of the interleukin 3 dependent myeloid cell line FDC-P1. Mol Cell Biol 7: 1673–1630

Wilkinson DG, Bailes JA, McMahon AP (1987) Expression of the proto-oncogene int-1 is restricted to specific neural cells in the developing embryo. Cell 50: 79–88

Willman CL, Stewart CC, Lin H, Tomasi TB (1987) Modulation of c-fms and c-fgr expression in normal monocytic cells stimulated to proliferate with CSF-1 monocytic colony-stimulating factor. J Cell Biochem [Suppl] 11 A: 201

Willumsen BM, Papageorge AG, Kung HF, Bekesi E, Johnsen M, Vass WC, Lowy DR (1986) Mutational analysis of a ras catalytic domain. Mol Cell Biol 6: 2646–2654

Witte ON (1986) Functions of the abl oncogene. Cancer Surv 5: 183–197

Wome RB, Frick K, Mitchell CD, Ross AH, Bishayee S, Scher CD (1987) PDGF induces c-myc mRNA expression in MG 63 human osteosarcoma cells but does not stimulate cell replication. J Cell Physiol 132: 65–72

Wong PM, Chung SW, Raefsky E, Eaves CJ, Nienhuis AW (1988) Blast colonies containing hemopoietic progenitor cells can give rise to Abelson virus (A-MuLV)-transformed cell lines. Exp Hematol 16: 5–11

Wyke JA, Stoker AW, Searle S, Spooncer E, Simmons P, Dexter TM (1986) Perturbed hemopoiesis and the generation of multipotential stem cell clones in src-infected bone marrow cultures is an indirect or transient effect of the oncogene. Mol Cell Biol 6: 959–963

Yaffe D (1968) Retention of differentiation potentialities during prolonged cultivation of myogenic cells. Proc Natl Acad Sci USA 61: 477–483

Yamamoto T, Ikawa S, Akiyama T, Semba K, Nomura N, Miyajima N, Saito T, Toyoshima K (1986) Similarity of protein encoded by the human c-erbB2 gene to the epidermal growth factor receptor. Nature 391: 230–234

Yamanashi Y, Fukushige S, Semba K, Sukegawa J, Miyajima N, Matsubara K, Yamamoto T, Toyoshima K (1987) The yes-related cellular gene lyn encodes a possible tyrosine kinase similar to p56lck. Mol Cell Biol 7: 237–243

Yarden Y, Kuang W, Yang-Feng T, Coussens L, Munemitsu S, Dull TJ, Ellson C, Schlessinger J, Francke U, Ullrich A (1987) Human proto-oncogene c-kit: a new cell surface receptor tyrosine kinase for an unidentified ligand. EMBO J 6: 3341–3351

Yokota J, Tsunetsugu Yokota Y, Battifora H, Le Fevre C, Cline MJ (1986) Alterations of myc, myb, and ras Ha proto-oncogenes in cancers are frequent and show clinical correlation. Science 231: 261–265

Yuspa SH, Kilkenny AE, Stanley J, Lichti U (1985) Keratinocytes blocked in phorbol ester-responsive early stage of terminal differentiation by sarcoma viruses. Nature 314: 459–462

Zimmerman KA, Yancopoulos GD, Collum RG, Smith RK, Denis KA, Nau MM, Witte ON, Toran Allerand D, Gee CE (1986) Differential expression of myc family genes during murine development. Nature 319: 780–783

Genetics of *src*: Structure and Functional Organization of a Protein Tyrosine Kinase

J. Thomas Parsons and Michael J. Weber

Department of Microbiology and University of Virginia Cancer Center, University of Virginia School of Medicine, Charlottesville, VA 22908, USA

Current Topics in Microbiology and Immunology, Vol. 147

1 Introduction

> A tumor of the chicken, histologically a spindle-celled sarcoma, has been propagated in this laboratory since October 1909, and in the past few months has developed extreme malignancy. From a bit inoculated into the breast muscle of a susceptible fowl there develops rapidly a large, firm growth; metastasis takes place to the viscera; and within four to five weeks often the host dies. The behavior of the new growth has been throughout that of a true neoplasm, for which reason the fact of its transmission by means of a cell-free filtrate assumes exceptional importance.
>
> P. Rous (1910)

The observations of Peyton Rous almost 80 years ago provided the first clues to the viral etiology of cancer. Since that time the study of transforming retroviruses such as Rous sarcoma virus (RSV) and the subsequent identification of the genes responsible for the induction of the neoplastic state (oncogenes) have provided a foundation for understanding the molecular basis of cancer (Bishop 1987). Of equal importance is the recognition that transforming retroviruses transduce cellular gene sequences that encode proteins which function in the process of normal cell growth and differentiation. The analysis of transforming retroviruses therefore provides a unique opportunity to identify, isolate, and characterize cellular proteins that play fundamental roles in the regulation of normal and malignant cell growth. Genetic and biochemical studies using RSV have played a preeminent role in defining the relationship between virus expression and cellular transformation. Genetic evidence for a unique gene required for transformation was provided by Martin (1970), Kawai and Hanafusa (1971), and Toyoshima and Vogt (1969) with the isolation of temperature-sensitive mutants of RSV, thus clearly implicating a viral gene product in the transformation process. The demonstration that a unique viral sequence was required for cell transformation was provided by Vogt (1971) and Martin and Duesberg (1972) when it was shown that transformation-defective (td) variants of RSV lacked a segment of the viral genome, yet replicated efficiently in cells in culture. The isolation of sequences complementary to this unique viral sequence, originally termed *src* (v-*src*), led to the seminal observation that viral transforming sequences were present not only in the genome of the virus but also in the chromosomes of normal cells (c-*src*) (Stehelin et al. 1976). Next, the identification of the *src* gene product, pp60^{src} (Brugge and Erikson 1977), and the demonstration that pp60^{src} was a unique protein kinase (Collett and Erikson 1978; Levinson et al. 1978) with specificity for tyrosine phosphorylation (Hunter and Sefton 1980) opened the door to a wealth of biochemical and genetic experimentation on the role of tyrosine kinase oncogenes and protooncogenes in normal and malignant cells.

The purpose of this review is to summarize recent genetic studies of the *src* gene and discuss how genetic analysis of pp60^{src} has provided insights into both the structure and function of the viral and cellular *src* gene products, as well as the nature of their cellular substrates.

2 The *src* Protein

The *src* gene of RSV encodes à 60000-dalton phosphoprotein (pp60^{src}) first identified by Brugge and Erikson (1977) using sera from rabbits bearing RSV-induced tumors. pp60^{src} is a membrane-associated phosphoprotein (Courtneidge et al. 1980; Krueger et al. 1983) that exhibits tyrosine-specific protein kinase activity both in vitro and in vivo (Jove and Hanafusa 1987). The immunoprecipitation of pp60^{src} from extracts of RSV-infected cells using either tumor-bearing rabbit serum (TBR) or monoclonal antibody directed against the *src* protein yields immune complexes which, in the presence of Mg^{2+} (or Mn^{2+}) and ATP, catalyze both the phosphorylation of pp60^{src} itself (autophosphorylation) (Purchio 1982) and exogenous substrates (tyrosine-containing peptides or proteins) (Cooper et al. 1984a). That the tyrosine-specific protein kinase activity of pp60^{src} is required for initiation and maintenance of the transformed state can be deduced from two experimental observations. First, transformation by RSV is accompanied by a dramatic increase in the level of cellular phosphotyrosine (Sefton et al. 1980). Secondly, mutations in the *src* gene that inactivate the protein kinase activity of pp60^{src} are invariably defective for transformation (Bryant and Parsons 1984; Kamps and Sefton 1986; Snyder et al. 1985). These argue convincingly that the tyrosine kinase activity of pp60^{src} is indispensable for transformation; however, the mechanism by which tyrosine phosphorylation mediates the events leading to cellular transformation remains uncertain.

pp60$^{v\text{-}src}$ contains two major sites of phosphorylation (Fig. 1), Tyr-416 — the site of autophosphorylation (Patschinsky et al. 1982; Smart et al. 1981) — and Ser-17, a site of phosphorylation by cAMP-dependent protein kinase (Collett et al. 1979; Patschinsky et al. 1986). In addition, pp60^{src} can be phosphorylated on Ser-12 by protein kinase C (PKC), either in vitro by incubation with purified PKC or in

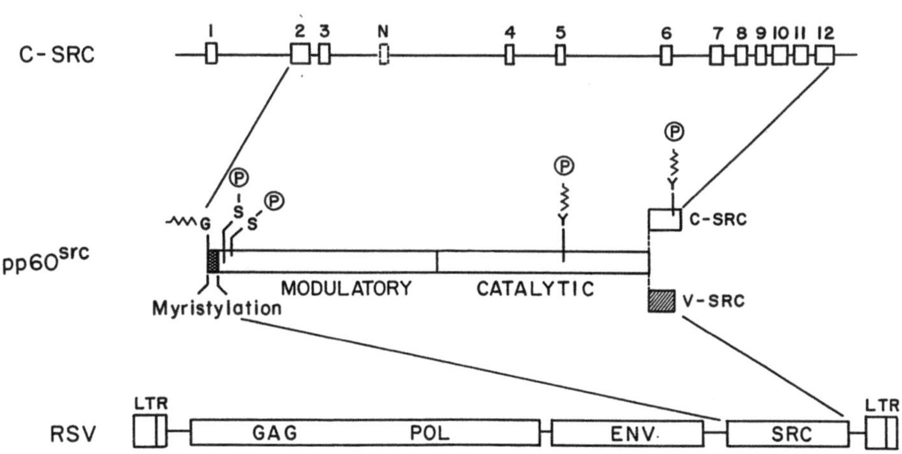

Fig. 1. Structure of the viral and cellular *src* gene. Schematic drawing shows the origin of cellular c-*src* coding exons (*top*) and viral *src* sequences (*bottom*). *Dashed box* (labeled *N*) in the c-*src* gene indicates the alternatively spliced exon present in neuronal c-*src*

vivo by treatment of cells with phorbol esters and other agents that activate PKC (Purchio et al. 1985; Gould et al. 1985). The significance of serine phosphorylation of pp60^{src} remains obscure. However, some changes in serine phosphorylation may have a modest effect on kinase activity (Roth et al. 1983; Gould and Hunter 1988). Moreover, the observation that other tyrosine kinases (both oncogene products and receptor kinases) are phosphorylated on serine and threonine leads one to speculate that serine phosphorylation is an important mechanism for modulating the interaction of *src* with other cellular kinases which, in turn, may play a role in the regulation of normal cellular *src* activity.

Shortly after synthesis, the viral *src* protein is found transiently associated with two cellular proteins, hsp90 and p50 (Brugge et al. 1981; Oppermann et al. 1981). It has been suggested that this association plays a role in transport of pp60^{src} to cellular membranes (Brugge 1986a). However, many thermosensitive *src* mutants display increased levels of the complex with p90 and p50, raising the possibility that p90 and p50 bind to a variety of partially denatured forms of pp60$^{v\text{-}src}$. This suggestion is consistent with the identification of hsp90 as a heat-shock protein (Brugge 1986).

pp60^{src} is found associated with virtually all cellular membranes (Courtneidge et al. 1980; Courtneidge and Bishop 1982; Krueger et al. 1983), with approximately 50% of the steady-state *src* protein being associated with plasma membranes (Resh and Erikson 1985b). The membrane association of pp60^{src} requires a 14-carbon saturated fatty acid, myristic acid, linked via an amide bond to the α-amino group of the amino-terminal glycine residue (Buss et al. 1984; Buss and Sefton 1985; Schultz et al. 1985). Myristylation of pp60^{src} appears to occur in a co-translational fashion (Wilcox et al. 1987; Deichaite et al. 1988) during the synthesis of pp60^{src} on cytoplasmic polyribosomes (Buss et al. 1984).

The locations at which pp60$^{v\text{-}src}$ is detected depends very much on the reagents and techniques used for the analysis. For example, antiserum raised against the carboxy-terminal half of the protein strongly displays pp60$^{v\text{-}src}$ in a perinuclear location, whereas more general reagents do not (Resh and Erikson 1985a). Substantial concentrations of pp60$^{v\text{-}src}$ can also be found localized in adhesion plaques, which represent the sites of termination of microfilament bundles and of close apposition of the cell to its substrate (Rohrschneider 1980). Extraction of whole cells with nonionic detergents releases lipids and cytosolic proteins, leaving a detergent-insoluble fraction which is greatly enriched in cytoskeletal proteins. Transformation-competent pp60^{src} is found largely in this cytoskeletal fraction after detergent extraction (Burr et al. 1980), whereas pp60$^{c\text{-}src}$ and nontransforming *src* mutant proteins are solubilized by detergents (Hamaguchi and Hanafusa 1987). Interestingly, the small amount of transformation-competent pp60$^{v\text{-}src}$ which is solubilized by nonionic detergent is hypophosphorylated (Comoglio et al. 1984). Although it is generally assumed that the principal activity of the *src* protein involves its association with plasma membranes (Willingham et al. 1979), it is quite possible that association with nonplasma membranes, adhesion plaques, or the cytoskeleton are important sites of *src* activity.

The cellular homologue of the viral *src* gene product (pp60$^{c\text{-}src}$) is encoded by a single chromosomal locus (Takeya and Hanafusa 1983), and differs from its viral counterpart in several important respects (Fig. 1, 2). The cellular *src*

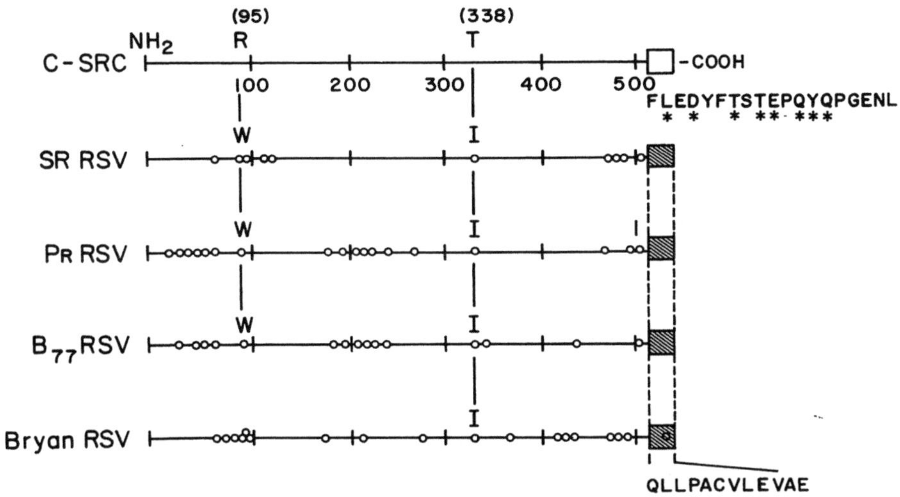

Fig. 2. Comparison of the predicted amino acid sequences of cellular and viral *src* proteins. The *open circles* denote amino acid changes present in the $pp60^{src}$ encoded by individual strains of RSV. Residues within the carboxy-terminal portion of $pp60^{c\text{-}src}$ that are conserved in all the members of the *src* family are indicated by *asterisks*

protein contains 19 carboxy-terminal amino acids that are distinct from the 12 carboxy-terminal residues of the virally encoded protein (TAKEYA and HANAFUSA 1983). Sequence comparison of c-*src* with different strains of RSV reveals, in addition, numerous mutations scattered throughout the entire sequence of the viral gene products (Fig. 2). Three strains of RSV, the Prague (Pr), Schmidt-Ruppin (SR), and B77, have in common amino acid substitutions at residues 95 and 338, whereas the Bryan (Br) strain shares the change at position 338 but not at position 95 (MAYER et al. 1986). Finally, expression of the avian c-*src* gene in rodent and avian cells has shown that $pp60^{c\text{-}src}$ is nononcogenic (SHALLOWAY et al. 1984; IBA et al. 1984; PARKER et al. 1984). These observations clearly indicate that alterations in the structure of the virally encoded *src* protein contribute to the oncogenicity of the RSV gene product.

3 *src* Transformation: Phenotypic Alterations

Although Peyton Rous isolated the RSV on the basis of its ability to cause tumors, most studies on the biological effects of *src* have exploited the fact that *src* and other oncogenes induce in cultured cells a variety of morphological, physiological, and growth-regulatory changes which are termed "the transformed phenotype." In general, these changes correlate with in vivo tumorigenicity, although individual exceptions have been noted (KAHN et al. 1982). The effects of *src* on fibroblastic cells have been studied in most detail, since these cells are easily cultured and presumably represent the primary targets for in vivo tumorigenicity by RSV. The effects of *src* are quite different in other cellular backgrounds; for example, *src* stimulates cellular differentiation rather than transformation in the PC12 pheochromocytoma cell line (ALEMA et al. 1985), and *src* mutants which are transformation-defective for fibroblasts are capable of stimulating the growth of neuroretinal cells (CALOTHY et al. 1978).

3.1 Morphological Phenotypes

The earliest manifestation of *src* expression is membrane ruffling along the cell margins and on the dorsal surface of the cell (AMBROS et al. 1975; WANG and GOLDBERG 1976). This change in membrane activity can be seen within minutes when cells infected with a temperature—conditional mutant are shifted from the restrictive to the permissive temperature, even in the presence of cycloheximide. It may thus represent a primary effect of pp60$^{v\text{-}src}$ activity.

Normal, cultured fibroblasts grow as monolayers of extended, flattened cells and in organized, parallel arrays. By contrast, transformed cells acquire a rounded, refractile morphology, and tend to pile up in disorganized clusters. Associated with the gross change in cellular morphology is a pronounced disorganization of the microfilament network which constitutes much of the cytoskeleton (WANG and GOLDBERG 1976). A defined set of *src* mutants induces a morphological change which is refractile, but long and highly extended — almost needle-like — rather than rounded. Such a phenotype was first described by TEMIN (1960) and termed "fusiform." The fusiform phenotype is readily apparent, and is distinct from an "intermediately transformed" phenotype, since it cannot be induced by varying quantitatively the amount of pp60$^{v\text{-}src}$ activity (e.g., by placing cells infected with thermosensitive mutants at various intermediate temperatures) (WEBER and FRIIS 1979).

3.2 Growth Regulation Phenotypes

Changes in growth control are the *sine qua non* of malignant transformation. Cells transformed by *src* require reduced levels of serum growth factors, and are capable of growing to a higher cell density. They thus have lost "density-dependent" growth regulation. The ability to overgrow, combined with the morphological changes described above, constitute the basis for the "focus assay" which is used to quantitate the number of functional RSV particles. Fibroblasts ordinarily require anchorage to a matrix or solid support for growth. Transformation abrogates this requirement, so that *src*-transformed cells are capable of growing in suspension or in a semi-solid medium ("anchorage-independent growth").

3.3 Physiological and Molecular Phenotypes

A number of changes in gene expression have been noted in *src*-transformed cells (BRZESKI and EGE 1980; GROUDINE and WEINTRAUB 1975; HENDRICKS and

WEINTRAUB 1984) including increased synthesis of plasminogen activator (RIFKIN et al. 1975) and increased expression of a gene coding for a potential growth-regulatory factor (SUGANO et al. 1987; BEDARD et al. 1987). Decreased expression of the genes for α_2-collagen (ALLEBACH et al. 1985; DE CROMBRUGGHE 1985), the epidermal growth factor (EGF) receptor (WASILENKO et al. 1987; W. J. WASILENKO and M. J. WEBER, unpublished observations) and fibronectin (TYAGI et al. 1985) have also been reported. The changes in fibronectin gene expression are particularly noteworthy, since the fusiform mutants appear to be specifically defective in the induction of this phenotype (ROHRSCHNEIDER and REYNOLDS 1985) and addition of fibronectin to transformed cells partially reverses morphological transformation (ALI et al. 1977).

Increased transcription of the glucose transporter gene has been reported in *src*-transformed rodent fibroblasts (FLIER et al. 1987) but, surprisingly, no changes in transporter transcription have been detected in transformed chicken embryo fibroblasts. In the avian cells, the *src* protein increases the amounts of transporter by inducing a decrease in the rate at which the protein is degraded. Thus, *src* is able to induce the same phenotypic end-point by different molecular mechanisms in the two different cell types (WHITE and WEBER 1988).

Decreases in intercellular communication (AZARNIA et al. 1988; ATKINSON and SHERIDAN 1985; CHANG et al. 1985), cAMP levels and adenylate cyclase activity (ANDERSON and PASTAN 1975) occur rapidly following *src* expression and, like the morphological changes, may represent primary consequences of $pp60^{v\text{-}src}$ activity. The change in cAMP levels may play a role in controlling changes in cell shape, growth, and glucose transport, since addition of cAMP analogs to transformed cultures tends to restore these parameters to a more normal condition (WEBER 1974).

4 Genetic Analysis of The Viral *src* Protein

4.1 Evidence for a Domain Structure of the *src* Protein

The overall structure of the *src* protein can be viewed as a series of interactive structural and functional domains[1] (Fig. 3), the boundaries of which have been defined by biochemical analysis, amino acid sequence comparisons, and genetic analysis. The catalytic activity of $pp60^{src}$ resides within a carboxy-terminal 30-kDa proteolytic fragment that exhibits tyrosine kinase activity in vitro (BRUGGE and DARROW 1984). This region of the *src* protein also shares amino acid sequence homology with other members of the "*src*" protein kinase family, other receptor tyrosine kinases, and, to a limited extent, other serine/threonine kinases (for reviews see HUNTER and COOPER 1985; HANKS et al. 1988). Mutations within the catalytic domain usually have effects on kinase activity and transforming potential (see below).

[1] The authors note that the term "domain" denotes a region of the protein defined by the linear sequence of the *src* protein and is used in its most general sense!

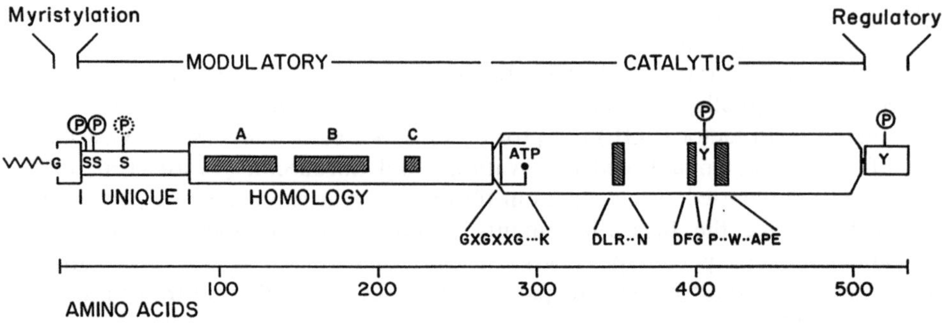

Fig. 3. Model depicting the organization of domains in pp60$^{c\text{-}src}$. The domain structure of pp60$^{c\text{-}src}$ is predicted based on sequence comparisons and interpretation of genetic data (see text). The *dashed boxes* within the homology region denote the regions of reported sequence similarity between phospholipase C 148 and *src* (STAHL et al. 1988, SUH et al. 1988). *Dashed boxes* within the catalytic domain denote highly conserved amino acid sequence motifs present in members of the tyrosine kinase family (see text). Sites of serine and tyrosine phosphorylation are indicated as Ⓟ; a minor site of serine phosphorylation is denoted by Ⓟ. Single-letter amino acid designations: *A*, Ala; *G*, Gly; *D*, Asp; *F*, Phe; *K*, Lys; *W*, Trp; *T*, Thr; *S*, Ser; *Y*, Tyr; *P*, Pro; *N*, Asn; *L*, Leu; *R*, Arg

Myristylation of pp60^{src} and subsequent membrane binding is mediated via an amino terminal domain encompassing the first seven to ten amino acids of pp60^{src}. As discussed in detail below, this domain contains the site of myristylation and additional residues required for efficient recognition by the myristylation enzymes.

The remaining amino terminal sequences comprise what can be termed the "modulatory region" of the *src* protein (JOVE and HANAFUSA 1987). This region contains major sites of phosphorylation by other protein kinases (Ser-12, Ser-17) as well as regions that exert negative and positive effects on the activity of pp60^{src} (see below). The region encompassing residues 18–83 is quite divergent within the *src* family of tyrosine kinases, and has been termed the "unique" domain (COOPER 1989). Beginning at about residue 84, significant sequence similarity is evident among the nonreceptor tyrosine kinases and extends to the catalytic domain. This region (referred to here as the homology region) contains the region originally designated SH2 (*src* homology 2) based on sequence comparisons of the v-*fps*, v-*abl* and *src* genes (SADOWSKI et al. 1986). However, this region also exhibits striking sequence similarity with the protein product of *crk*, an oncogene which does not encode a tyrosine kinase (MAYER et al. 1988), phospholipase C (PLC) 148 (STAHL et al. 1988; SUH et al. 1988), and the GTP-activating protein, GAP (VOGEL et al. 1988) (Fig. 3). As discussed below, this region can also be subdivided based on the biological properties of mutations within the region and the homologies with *crk*, PLC 148, and GAP.

The product of the cellular *src* gene, pp60$^{c\text{-}src}$, is negatively regulated in vivo. This regulation is mediated by a carboxy-terminal domain comprising 19 amino acids missing from the v-*src* protein (COOPER et al. 1986; COURTNEIDGE 1985). Deletion of this domain or mutation of the major site of tyrosine phosphorylation, Tyr-527, results in activation of tyrosine kinase activity and concomitant activation of transforming potential (CARTWRIGHT et al. 1987a; KMIECIK and SHALLOWAY 1987; PIWNICA-WORMS et al. 1987; REYNOLDS et al. 1987). The carboxy-terminal regulatory domain is highly conserved among the *src* kinase family (Fig. 2) and the deletion of these regulatory sequences is a hallmark of the tyrosine kinase oncogenes acquired by transforming retroviruses. This domain of *src* also contains the site of polyoma mT antigen binding (CHEN et al. 1988). Binding of mT antigen activates the kinase activity of $pp60^{c\text{-}src}$ and may play a role in the mT antigen-directed transformation of cells (BOLEN et al. 1984; CARTWRIGHT et al. 1985; COURTNEIDGE 1985; YONEMOTO et al. 1985; YONEMOTO et al. 1987).

To simplify the discussion of the many *src* mutants and in an attempt to correlate genetic and biochemical evidence, we will discuss individual mutants in the context of a putative domain structure for *src* with the knowledge that the interactive nature of these domains dictates the properties of individual *src* proteins.

4.2 Myristylation Domain

Association of $pp60^{src}$ with cellular membranes is required for transforming activity of v-*src* and, presumably, for the normal function of the cellular *src* protein (WILSON et al. 1989). Cell fractionation studies have shown that unlike most membrane proteins $pp60^{src}$ is synthesized on soluble polyribosomes (COURTNEIDGE and BISHOP 1982; BRUGGE 1986a). In addition, sequence analysis revealed that the *src* gene product lacks a conventional hydrophobic signal sequence (SCHWARTZ et al. 1983). The interaction of $pp60^{src}$ with cellular membranes requires the covalent modification of $pp60^{src}$ by addition of myristic acid to an amino-terminal glycine residue (BUSS et al. 1984; BUSS and SEFTON 1985; SCHULTZ et al. 1985; GARBER et al. 1985). The importance of myristylation has been clearly documented by mutational analysis of the amino-terminal sequences of $pp60^{v\text{-}src}$ (Table 1). Initial studies indicated that substitution of peptide sequences between amino acids 1 (Met) and 2 (Gly), or peptide substitutions for amino acids 2–4, 3–4, or 7–15 block transformation, myristylation, and membrane association (CROSS et al. 1984; PELLMAN et al. 1985b). In contrast, substitution of a peptide for amino acids 11–15 does not alter the transforming properties or the myristylation and membrane association of the mutated *src* protein (PELLMAN et al. 1985b), suggesting that sequences required for efficient myristylation reside within the first 10 residues of $pp60^{src}$. Site-directed mutation of Gly-2 to Ala or Glu prevents myristylation and membrane association and inhibits cellular transformation, confirming that Gly-2 is the major site of fatty acid modification (KAMPS et al. 1985; BUSS et al. 1986).

Gene fusion experiments have further defined the sequences required for efficient myristylation. Addition of the first 14 amino acids of $pp60^{src}$ to either the v-*fps* protein or the α-globin gene product is sufficient to direct the myristylation of these proteins (PELLMAN et al. 1985a). The fusion of the first seven residues of $pp60^{src}$ to pyruvate kinase is also sufficient to cause myristylation of this hybrid protein, indicating that the minimal recognition sequence may be contained within the first seven residues of $pp60^{src}$ (KAPLAN et al. 1988). Mutagenesis of amino-

Table 1. Mutations in the myristylation domain of pp60[src]

Mutant[a]	Sequence alteration[a]	Ref.	Morphology[b]	Anchorage-Independent Growth	Kinase activity[c]		Localization[d]	Comments
					In vitro	In vivo		
SD 10	G2 → A	1	td	—	E^+Ang^+	+	C	Infected cells grow to slightly higher cell density
SD 11	G2 → E		td	—	E^+Ang^+	+	C	
NY 314	*dl*2–81 (*is*DL)	2	td	—	IG^+	+	C	
NY 315	*dl*2–15 (*is*DLG)		td	—	IG^+	NR	C	
NY 300	*dl*2–4 (*is*NRSG)		td	—	IG^+	NR	C	
NY 304	M1 (*is*DL)	3	td	—	IG^+	NR	C	
NY 306	*dl*3–4 (*is*RSG)		td	—	IG^+	NR	C	
NY 316	*dl*7–15 (*is*NRSG)		td	—	IG^+	NR	C	
NY 317	*dl*11–15 (*is*QICG)		T	+	IG^+	NR	M	
SF N7	K7 → N	4	td	NR	IG^+	NR	C	
SF R7	K7 → R		T	NR	IG^+	NR	M	
rASV 157	Env (30aa) → S6	5	f/heterogenous	+	IG^+	+	salt-sensitive membrane association	Poorly tumorigenic, partially transforming
rASV 1702	Env (45aa) → A76		f/heterogenous	+	IG^+	+		

[a] Amino acid alterations are denoted by the single letter designation and/or the position of the sequence change within the *src* protein. *is* denotes insertion and *dl* deletion of amino acids. All NY, CU SHX, SN, STX, SRX, SDX, SAX, and SPX v-*src* mutants are derived from SR RSV; all CH, SD and LA mutants from Pr RSV. A, Ala; R, Arg; N, Asn; D, Asp; C, Cys; Q, Gln; E, Glu; G, Gly; H, His; I, Ile; L, Leu; K, Lys; M, Met; F, Phe, P, Pro; S, Ser; T, Thr; W, Trp; Y, Tyr; V, Val.

[b] The morphology of infected cells is denoted by: T, indistinguishable from wild-type RSV-transformed cells; td, no appreciable alteration in phenotype; f, fusiform morphology. For temperature-sensitive mutants, the temperature at which the phenotype was determined is shown in parentheses.

[c] IG denotes the phosphorylation of immunoglobulin G heavy chain; A, autophosphorylation of pp60[src]; E, enolase phosphorylation; Ang, angiotensin phosphorylation; + denotes increased tyrosine phosphorylation of cellular proteins measured by either phosphoamino analysis, alkali treatment of one- and two-dimensional polyacrylamide gels or antiphosphotyrosine immunoblotting.

[d] Intracellular localization of mutant pp60[src] is denoted by C, cytoplasmic, or M, membrane associated.

[e] NR not reported.

References: 1, Kamps et al. 1985, Kamps et al. 1986, Buss et al. 1986; 2, Cross et al. 1984; 3, Pellman et al. 1985b; 4, Kaplan et al. 1988; 5, Krueger et al. 1982a, b, 1984, Garber and Hanafusa 1987.

terminal sequences has provided evidence that Lys-7 is a critical residue in the recognition sequence required for efficient myristylation. Replacement of Lys-7 by Asn, but not by Arg, greatly reduced the myristylation, membrane association and transforming activity of pp60^{src} (KAPLAN et al. 1988).

Examples of alternative mechanisms of membrane association are provided by the analysis of two isolates of recovered avian sarcoma viruses (rASV) (KRUEGER et al. 1982a, 1984), viruses generated as a consequence of recombination between a transformation-defective RSV (*src* deletion) and the cellular c-*src* gene. Two such isolates (1702 and 157) partially transform cells in culture and exhibit reduced tumorigenicity in vivo (Table 1). The *src* protein encoded by these mutants is unmyristylated and tends to accumulate in adhesion plaques. Membrane association occurs because of a novel deletion which results in the fusion of env signal peptide sequences to the amino-terminal sequence of *src* (Table 1) (GARBER and HANAFUSA 1987). The membrane and cytoskeletal association of these recovered *src* proteins is less stable than that of the wild-type pp60$^{v\text{-}src}$, being disrupted by treatment with high salt, and may be responsible for the reduced tumorigenicity and unique phenotypic properties of the variants (KRUEGER et al. 1982a, 1984). The fact that *src* variants with alternative mechanisms of membrane association are still capable of inducing some parameters of cellular transformation confirms that correct membrane localization per se plays an essential role in transformation.

Recent biochemical evidence suggests that myristylation of pp60^{src} may not be sufficient in itself for the stable and proper orientation of *src* protein in the membrane. RESH (1988), using in vitro membrane reconstitution experiments, has shown that stable association of *src* protein with membrane vesicles requires the myristylation of *src* and the presence of cellular proteins in the reconstitution mixture. These results argue for the involvement of cellular proteins in the stable insertion of pp60^{src} in cellular membranes.

Analysis of several mutant *src* proteins indicates that other structural features of pp60^{src} influence its stable association with cellular membranes. For example, the *src* protein encoded by NY 18-3 which lacks residues 149–264 and appears to be fully myristylated, does not associate with cellular membranes (Table 2). This lack of membrane association of NY 18-3 *src* protein correlates with increased association with p50 and p90 proteins (GARBER et al. 1985). Mutations within the catalytic domain of the *src* protein have also been shown to influence membrane association. A number of temperature-sensitive (ts) mutants, for example LA 29, exhibit a reduction in membrane binding particularly at the nonpermissive temperature, 41°, where virtually all of the pp60^{src} is cytosolic (STOKER et al. 1986) (Table 3). However, at both the permissive or nonpermissive temperatures, the mutant *src* protein appears fully myristylated (STOKER et al. 1986). Since the critical mutation in LA 29 is at position 507, it has been suggested that membrane association at the nonpermissive temperature is influenced by a defect in transit to the membrane (STOKER et al. 1986; WELHAM and WYKE 1988). Alternatively, the inability of the mutant *src* protein at the nonpermissive temperature to phosphorylate a critical protein may contribute to its membrane instability.

In summary, both genetic and biochemical experiments support the notion that myristylation of the *src* protein requires a domain encompassing the first seven residues of pp60^{src}. However, the association of myristylated pp60^{src} with the mem-

Table 2. Mutations in the modulatory region of $pp60^{src}$

Mutant[a]	Sequence alteration	Ref.	Morphology[b]	Anchorage-independent growth	Kinase activity[c]		Localization[d]	Comments
					In vitro	In vivo		
Unique domain								
NY 11-1	*dl*15 (*is*PRSG)		T	+	IG^+	+	M	Mutant $pp60^{src}$ contains significantly reduced Ser-17 phosphorylation
NY 11-4	*dl* 15 (*is*PQICG)	1	T	+	IG^+	+	M	
NY 11-7	*dl*15–27 (*is*PQIW)		T	+	IG^+	+	M	
CH 17C	S17 → C	2	T	+	A^+	+	NR	
NY 308	*dl*15–49 *is*PRSG	3	T	+	IG^+	+	M	
NY 309	*dl*15–81 *is*PDL		f	+	IG^+	+	M	
CH 128	*dl*53–112 *is*QAPSLELG	4	T	+	A^+E^+	+	M	
SHX 1	14 *is*RPRG		T	NR	IG^+	NR	NR	
SNX 2	52 *is*LE	5	T	NR	IG^+	NR	NR	
SHX 2	76 *is*DLEV		T	NR	IG^+	NR	NR	
Homology domain								
CH 129	*dl*37–171 *is*RGASR		td	–	A^+E^+	+	M	
CH 127	*dl*92–127 *is*PKRAW	4	T	+	A^+E^+	+	M	
CH 126	*dl*102–174 *is*PKRG		td	–	A^+E^+	+	M	
CH 130	*dl*111–143 *is*PKLAWA		T	+	A^+E^+	+	M	
STX 1	138 *is*RA	5	T	NR	IG^+	NR	NR	
SRX 1	148 *dl*Y, *is*SRD		T	NR	IG^+	NR	NR	
NY 311	*dl*15–149 *is*PRSD		f	+	IG^+	+	M	Soft agar colonies smaller than wild type (wt)
NY 310	*dl*15–169 *is*PRSG	6	td	–	IG^+	NR	M	Defective for soft agar growth
NY 312	*dl*15–264 *is*PRSG		td	–	IG^-	NR	M	
NY 320	*dl*149–169 *is*SQICG		td	–	IG^+	+	M	
NY 18-3	*dl*169–264 *is*PQICG		td	–	IG^-	NR		Sequestered as complex with p50:p90

CH 119	*dl*173–227	7	ts (41)	ts	A^+E^+	+	M	Kinase activity slightly reduced at nonpermissive temperature
CH 120	*dl*168–226 *is*PSM		ts (41)	ts	A^+E^+	+	M	
CH 125	*dl*111–197	4	td	–	A^+E^+	+	M	
CH 121	*dl*82–170 *is*DGG		td	–	A^+E^+	+	M	
CH 148	*dl*148–150		T	+	A^+E^+	+	M	
CH 155	*dl*155–157		td	–	A^+E^+	–	M	
CH 161	*dl*161–163		td	–	A^+E^+	–	M	
CH 165	*dl*165–168		td	–	A^+E^+	+	M	Transformation-defective *src* proteins exhibit increased rate of turnover
CH 169	*dl*169–171	8	td	–	A^+E^+	–	M	
CH 171	*dl*171–173		td	–	A^+E^+	–	M	
CH 175	*dl*175–177		td	–	A^+E^+	–	M	
CH $161LE_2$	L161 *is*LELE		td	–	A^+E^+	–	M	
CH 161LE	L161 *is*LE		T	+	A^+E^+	+	M	Slightly increased turnover of $pp60^{src}$
CU 12	E106 → K V461 → M	9	f	+	IG^+	+	diffuse	
*dl*5	*dl*135–236 *is*33 amino acids	10	f	+	IG^+	+	–	*src* protein absent from adhesion plaques
SF 000	frameshift at 146	11	td	–	IG^+	NR	NR	
SF/LO104	*dl*172	12	f	+	IG^+	+	M	Host range mutation. p60 absent from adhesion plaques
WO 101	*dl*116 → 141	13	f	+	IG^+	+	NR	
ST 529	*dl*116 → 141 K93 → E		f	+	IG (ts)	+	NR	

Table 2 continued

Mutant[a]	Sequence alteration	Ref.	Morphology	Anchorage-independent growth	Kinase activity[c]		Localization[d]	Comments
					In vitro	In vivo		
SHX 3	168 *is*RPRG		T	NR	IG^+	Nm	NR	Growth in agar, in vivo phosphorylation and membrane localization not reported for these mutants
SHX 4	174 *is*RPRG		T		IG^+			
SHX 3	168 *is*RPRG		T		IG^+			
SHX 4	174 *is*RPRG		T		IG^+			
SDX 2	203 *dl*I, *is*TRV		T		IG^+			
		5	f (41)					
SAX 2	206 *is*LE		T		IG^+			
SPX 1	225 *is*LE		f		IG (reduced)			
SPX 13	228 *is*RA		f		IG (reduced)			
STX 2	259 *is*RA		T		IG^+			
SHX 5	263 *is*RPRG		td		IG^-			
SAX 3	272 *is*LE		td		IG^-			

[a] See Table 1 for definition of symbols.

References: 1, Cross and Hanafusa 1983; 2, Reynolds, Bouton and Parsons, unpublished observations; 3, Cross et al. 1984; 4, Raymond and Parsons 1987; 5, DeClue and Martin 1989; 6, Cross et al. 1985; 7, Bryant and Parsons 1982; 8, Wang and Parsons 1989; 9, D. Wages, J. Keefer, A. Espeseth, T. Parsons and M. Weber, unpublished data; 10, Iwashita et al. 1983, Kitamura and Yoshida 1983; 11, Mardon and Varmus 1983; 12, Verderame et al. 1989; 13, Anderson and Fujita 1987.

Table 3. Mutations in the catalytic domain of $pp60^{src}$

Mutant[a]	Sequence alteration	Ref.	Morphology[b]	Anchorage-independent growth	Kinase activity[c]		Localization[d]	Comments
					In vitro	In vivo		
SD 1	K295 → E		td	—	Ang^-	—	NR	SD1 and SD2 rapidly revert in tissue culture
SD 2	K295 → R	1		—	Ang^-	—	NR	
SD 3	K295 → H		td	—	Ang^-	—	NR	
SF 2	K295 → M	2	td	—	IG^-	—	M	
SF 1	Y416 → F	3	T	+	IG^+	+	NR	Mutant exhibits reduced tumorigenicity in vivo
NY 10-1	*dl*412–416 *is*RSD	4	T	+	IG^+	+	NR	
CH 26	A430 → V		td	—	A^-, IG^-	—	M	
CH 9	P431 → S	5	td	—	A^-, IG^-	—	M	
CH 6	E432 → K		td	—	A^-, IG^-	—	M	
CH 1, 65	A433 → T		td	—	A^-, IG^-	—	M	
CH 1511	503 frameshift		td	—	A^-E^-	—	M/C	
CH 1545C	515 frameshift	6	td	—	A^-E^-	—	M/C	Stable membrane association reduced
CH 1545H	515–517 *is*PKLGPSLGV		td	—	A^-E^-	—	M/C	
SHX 14	293 *is*RA		td		IG^+			Growth in agar, in vivo phosphorylation, and membrane association not reported for this group of mutants
SSHX 6	299 *is*DLEV		ts (41)		IG^+			
SNX 3	300 *is*LE		ts (41)		IG^+			
SRX 4	356 *dl*Y, *is*SRD		T		IG^+			
		7						
SRX 5	416 *dl*Y, *is*SRD		ts (41)		IG^+			
SRX 6	509 *dl*Y, *is*SRD		td		IG^-			
SHXD 21	514 *is*RA		ts (41)		IG^+			
SPX 2	515 *is*LE		T		IG^+			

Table 3 continued

Mutant[a]	Sequence alteration	Ref.	Morphology[b]	Anchorage-independent growth	Kinase activity[c] In vitro	Kinase activity[c] In vivo	Localization[d]	Comments
CU 2	A403 → T D413 → G V461 → M	8	bleb	small	50%	50%	adhesion plaques	Infected cells have lowered glucose transport, blebby morphology, and form small colonies in soft agar
tsCU 11	G236 → A V461 → M T508 → I	8	td	+	50%	50%	M	
tsNY 68	*dl*351–354 V461 → M	9	ts (41)	ts	ts	ts	ts	All parameters of transformation ts. Both mutations required for temperature sensitivity
NY 72-4	*dl*164–166 → K F197 → F D208 → N V461 → M P503 → S	10	ts (41)	ts	ts	ts(low)	M & C (ts)	Only the two kinase domain mutations are necessary for ts phenotype
tsLA 29	K351 → R A375 → T P507 → A	11	ts	ts	ts	ts	ts	P507 mutation is sufficient for ts phenotype. Confers fusiform morphology
LA 29R	K427 → R +LA29 mutations		T	T	IG^+	+	M	

tsLA 32	R107 → P							R107 mutation confers altered morphology. G300 and R419 mutations confer ts phenotype
	G300 → V	11	ts (41)	ts	IG^{+}	+	M	
	R419 → Q				A ts			
tsLA 24	R480 → H		ts (41)	ts	ts	N.R.	M	
LA 24R	R480 → H	11	T	ts	+	N.R.	M	
	H492 → Y							
tsLA 31	G478 → D		ts (41)	ts	ts	N.R.	M	
LA $31R_1$	G478 → D		T	+	+	N.R.	M	
	H492 → M	11						
LA $31R_2$	G478 → D		T	+	+	N.R.	M	
	V377 → M							
PA 101	A53 → T							Only kinase domain mutations are required for temperature sensitivity
	P58 → L							
	I85 → T	10	ts (41)	ts	low	ts	M + C (34 & 41)	
	G105 → R							
	V328 → M							
	V524 → I							
PA 104	L325 → P	10	ts (41)	ts	low	ts	M + C (34 & 41)	L325 mutation by itself confers some temperature sensitivity
	V461 → M							

[a] See Table 1 for definition of symbols.

References: 1, Kamps and Sefton 1986; 2, Snyder et al. 1985, Snyder and Bishop 1984; 3, Snyder et al. 1983; 4, Cross and Hanafusa 1983; 5, Bryant and Parsons 1984; 6, Wilkerson et al. 1985; 7, DeClue and Martin 1988; 8, D. Wages, J. Keefer, A. Espeseth, T. Parsons, and M. Weber, unpublished data; 9, Nishizawa et al. 1985, Garber et al. 1983; 10, Mayer et al. 1986, Jove et al. 1986a, b, Garber et al. 1987; 11, Fincham et al. 1982, Fincham and Wyke 1986, Stoker et al. 1984, 1986, Wyke and Stoker 1987, Welham and Wyke 1988.

brane is certainly more complex, and likely requires, in addition, the interaction of several other domains of pp60^{src} with one or more cellular (membrane?) components.

4.3 Modulatory Region

4.3.1 Serine Phosphorylation Domain

pp60^{src} is phosphorylated at residue 17 by cAMP-dependent protein kinase (Collett et al. 1979; Patschinsky et al. 1986) and at residue 12 (major) and residue 48 (minor) by protein kinase C (PKC; Purchio et al. 1985; Gould et al. 1985). Although these protein kinases are thought to be important regulatory kinases in the signal-transduction process, their role in the regulation of pp60^{src} activity is uncertain. The deletion of Ser-17 in v-*src* (Cross and Hanafusa 1983) or the mutation of Ser-17 to Cys or Tyr (A. Reynolds and T. Parsons, unpublished data) does not influence cell transformation. Similarly, mutation of Ser-12 of v-*src* has little effect on the transforming potential of this variant in chicken cells (A. Reynolds, A. Bouton, and T. Parsons, unpublished data). Therefore, the posttranslational modification of v-*src* at either Ser-17 or Ser-12 appears to have little influence on morphological transformation. Direct analysis of pp60$^{c\text{-}src}$ has shown that tumor promoters as well as agents that activate phosphatidylinositol turnover [platelet-derived growth factor (PDGF) fibroblast growth factor (FGF) serum, vasopressin, and prostaglandin $F_{2\alpha}$] stimulate the phosphorylation of pp60^{src} on Ser-12. In addition, administration of PDGF to quiescent fibroblasts induces the phosphorylation of one or two additional serine residues and one tyrosine residue within the amino-terminal 16 kDa of c-*src* and activates the immune complex kinase activity two- to threefold (Ralston and Bishop 1985; Gould and Hunter 1988). Hirota et al. (1988) report that the mutation of Ser-12 of the c-*src* gene leads to increased tyrosine kinase activity and unusual phenotypic effects. The high-level expression of a c-*src* variant containing a Ser-12 to Cys or Tyr alteration results in unusual morphological alterations of Rat 1 cells and increased anchorage-independent growth (A. Bouton, A. Reynolds, and T. Parsons, unpublished data). While these data argue for a possible role of phosphorylation in regulation of pp60^{src} activity, the precise nature and function of this complex regulatory network remain to be elucidated.

4.3.2 Unique Domain

This domain encompasses amino acid residues 18–84 and represents the region of significant amino acid sequence diversity among the *src* family of tyrosine kinases (Cooper 1989) (Fig. 3). Although this domain contains at least one site of serine phosphorylation (Ser-48), deletions within this region of v-*src* appear to have little if any effect on transformation by v-*src*. Raymond and Parsons (1987) reported that mutants containing deletions within this region of the Pr RSV are capable of inducing all parameters of transformation (Table 2). Deletions within this region of the SR RSV strain yield variants that are indistinguishable from wild type (*dl*15–49) or, in the case of one large deletion, induce a fusiform morphology (*dl*15–81) (Cross et al. 1985). DeClue and Martin (1989) have shown that linker insertions within this region of SR RSV do not alter phenotypic transformation of chicken or rat cells. It is interesting to speculate that this region of *src* (and perhaps the equivalent region in the other *src*-related tyrosine kinases) directs the interaction of c-*src* with a unique cellular "effector" molecule. This interaction would be abrogated and unnecessary in the case of v-*src*, a constitutive tyrosine protein kinase. As we will discuss below, this model gains further support from the analysis of c-*src* mutations.

4.3.3 Homology Domain

Analysis of both site-directed mutations and naturally occurring mutations indicates that the homology domain of *src* plays a significant role in modulating the function of both v-*src* and c-*src* encoded proteins. In fact, properties of individual mutants with structural alterations within this region, and the similarity of sequences within this domain with PLC, the *crk* oncogene, and the GAP protein suggest that the homology region can be further subdivided into at least three regions, A, B, and C (Fig. 3).

Structural alterations within the A region appear to have different effects, depending on the strain of RSV. Raymond and Parsons (1987) reported that deletions within the region defined by amino acids 84–149 of PrA RSV [CH 128 (*dl*53–112), CH 130 (*dl*111–143)] had little effect on the transformation of chicken cells (Table 2); however, deletions 3′ to this region were defective for transformation [CH 126 (*dl*102–174)]. Similarly, deletions within an analogous region of SR RSV [NY 311 (*dl*15–149)] induced a fusiform morphology, whereas deletions 3′ to this region [NY 310 (*dl*15–169)] were significantly more defective in inducing morphological changes and were severely impaired in their ability to induce anchorage-independent growth (Cross et al. 1985).

There is increasing evidence that the B and C regions (defined here as encompassing residues 149–260) (Fig. 3) direct the interaction of the *src* protein with cellular factors required for the initiation and maintenance of transformation. Mutations within this region have unusual transforming properties and confer, in some instances, a "host range" phenotype, i.e., the ability to transform one cell type (avian) but not another (rat). Bryant and Parsons (1982) first reported that site-directed deletion of amino acids 173 to 227 yielded a variant that was ts for transformation (tsCH 119) and encoded a *src* protein with only slightly reduced tyrosine kinase activity. Using site-directed deletion techniques, it was subsequently shown that other mutations within this region lead to alterations in morphology and growth properties (Table 3). A SR RSV variant lacking amino acids 149–169 (NY 320) induced only subtle morphological changes and was severely compromised for anchorage-independent growth (Cross et al. 1985) (Table 3). A larger deletion within this region (residues 169–264, NY 18-3) yielded a transformation-defective variant which displayed significantly reduced tyrosine protein kinase activity (Cross et al. 1985). Using the Pr strain of RSV, Raymond and Parsons (1987) showed that large deletions impinging on the B region of *src* were transformation-defective [CH 125 (*dl*111–197), CH 126 (*dl*102–174)]. Wang and Parsons (1989) extended these studies to show that a series of tri- and tetra-amino acid deletions through this region altered *src* transforming activity. Whereas deletion of amino acids 145–148 yielded a variant indistinguishable from wild-type RSV, deletion of residues 155–157, 161–163, 165–168, 169–171, 171–173, or 175–177 blocked the efficient transformation of chicken embryo cells (Table 3). The *src* proteins encoded by these transformation-defective variants showed an increased rate of turnover, but they retained the ability to associate with cellular membranes and their in vitro tyrosine kinase activity. DeClue and Martin (1989) have reported that insertion mutations at positions 152, 225, and 228 of SR-RSV *src* resulted in variants that induced a fusiform morphology, whereas an insertion at position 203 yielded a variant that induced wild-type transformation at 35° but a fusiform morphology at 41°. A larger deletion variant lacking residues 149–174 induced a fusiform morphology. In these latter studies, each of the variants that exhibited a fusiform morphology encoded *src* proteins with reduced tyrosine protein kinase activity, suggesting that the abundance and/or activity of the mutant proteins may have contributed to the altered phenotype.

Biologically isolated mutants of *src* mapped to the homology region induce a fusiform morphology (Table 2). For example, the fusiform mutant CU 12 contains amino acid substitutions at positions 106 and 461 (D. Wages, J. Keefer, A. Espeseth, T. Parsons, and M. Weber, unpublished data). Since the change at position 461 has been shown not to have biological effects by itself but appears to potentiate the effects of mutations at other sites (Nishizawa et al. 1985), the fusiform phenotype induced by this mutant is likely to be dependent upon the amino acid change at position 106. The SR RSV mutants SR 529, WO 101 and WO 201 (Fujita et al. 1981) all contain the same deletion (amino acids 116–141), and the mutant *dl*5 contains a deletion of residues 135–236 (with an insertion of 33 amino acids); these are also fusiform mutants (Iwashita et al. 1983; Kitamura and Yoshida 1983).

The SF/LO 104 mutant has a deletion of amino acid 172 and induces a fusiform morphology in chicken embryo fibroblasts but does not transform rat cells, even when expressed to levels which approach those obtained with wild-type v-*src* (Verderame et al. 1989). Thus, this mutation appears to exhibit a host-dependence for transformation. These findings are interpreted as indicating that this region of $pp60^{src}$ interacts with a critical cellular protein which differs subtly between rat and chicken cells, so that the mutant is able to interact efficiently with the normal (chicken) cellular protein, but is unable to interact appropriately in the rat species. Interestingly, the specific kinase activity of the mutant $pp60^{src}$ is lower in the rat than in the chicken cells, whether measured in vitro or by the in vivo tyrosine phosphorylation of cellular proteins (Verderame et al. 1989). Thus, the species-specific critical cellular protein may be a regulator of kinase activity rather than a substrate for the *src* kinase. It is not known whether the other biologically isolated fusiform mutants are also host-range restricted.

DeClue and Martin (1989) have reported that insertions within $pp60^{v\text{-}src}$ at positions 225 and 228 compromise the ability of such mutants to transform rat cells in culture. However, it remains to be established whether the mutant *src* proteins are efficiently expressed in these cells. Insertional mutagenesis of the v-*fps* oncogene has shown that a region of v-*fps* corresponding to the B-C domain of *src* (designated SH2) is required for efficient transformation of rat cells but not chicken cells (SADOWSKI et al. 1986; DECLUE et al. 1987). The identification of a host-range mutation in v-*fps* provides further evidence that the B-C or SH2 domain may regulate the interaction of tyrosine kinases with critical cellular proteins and that specific mutations alter this interaction in a species-specific fashion.

4.3.4 The Homology Domain — A Regulatory Domain?

The amino acid sequence of the homology region exhibits marked similarity with members of the *src* tyrosine kinase family, other membrane-associated tyrosine kinases and nontyrosine kinases, PLC-148, the v-*crk* oncogene, and GAP (Fig. 3). A functional link among these proteins is provided by the observation that the expression of the *gag-crk* oncogene elevates the level of tyrosine phosphorylation of cellular proteins, despite the fact that the v-*crk* protein is not a tyrosine protein kinase itself (MAYER et al. 1988). This observation has given rise to the suggestion that v-*crk* mediates transformation by altering the regulation of an existing membrane-associated tyrosine protein kinase. In light of the sequence homologies among these proteins, one can speculate that the B-C region might play an important role in directing the *src* protein to an important microenvironment within the membrane where it can interact with (and possibly phosphorylate) components of the signal transduction pathway. The analysis of phosphotyrosine-containing proteins in cells infected with homology domain variants should provide useful and interesting information regarding the specificity of tyrosine phosphorylation of cellular targets. Analysis of such phosphorylations is described later in this review.

In summary, properties of mutants within the homology domain indicate that the deletion of sequences within this region has significant effects on the capacity of the mutant *src* proteins to induce morphological alterations and changes in growth properties. It is unclear to what extent these effects are a result of the alterations in the association of $pp60^{src}$ with critical cellular (membrane) components or whether altering the structure of $pp60^{src}$ in this domain influences the intrinsic tyrosine kinase activity of $pp60^{src}$, thus altering its ability to phosphorylate specific cellular proteins.

4.4 Kinase Domain

The catalytic domain of the *src* tyrosine kinase is contained within an approximately 30-kDa region of the *src* protein. The boundaries of this domain as well as the functional importance of certain highly conserved regions within it can be deduced from sequence comparisons with other tyrosine and serine/threonine protein kinases (HUNTER and COOPER 1985; HANKS et al. 1988). Mutations within the catalytic domain of the *src* protein (residues 270–516) result in both temperature-sensitive and transformation-defective phenotypes. The relationships among structure, function, and phenotype are best considered within the context of what we currently understand about the function of conserved sequence motifs within the catalytic domain of protein kinases.

The consensus sequence Gly-X-Gly-X-X-Gly (residues 274–279) (Fig. 3) is found in many nucleotide-binding proteins in addition to protein kinases (WIERENGA and HOL 1983; HANKS et al. 1988), and in the tyrosine-specific kinases is flanked by invariant Leu and Val residues (positions 273, 281). Based on information gained from the analysis of other nucleotide binding proteins, a model for the ATP-binding site of pp60^{src} places the Gly-X-Gly-X-X-Gly sequence within an elbow around the nucleotide triphosphate with the first Gly residue in contact with the ribose ring and the second Gly residue near the terminal pyrophosphate (STERNBERG and TAYLOR 1984). A mutation adjacent to this region created by linker insertion of a dipeptide at residue 272 (SAX 3, Table 3) blocks transformation and yields pp60^{src} with greatly reduced kinase activity (DECLUE and MARTIN 1988). The invariant Lys residue at position 295 appears to be directly involved in the phosphotransfer reaction (KAMPS et al. 1984). This residue in pp60^{src} reacts with the ATP analog *p*-fluorosulfonyl-5′-benzoyl adenosine, an ATP analog that inhibits kinase activity (KAMPS et al. 1984). Site-directed mutations of this residue which change Lys-295 to Met, Glu, Arg, or His block transformation and kinase activity (SNYDER et al. 1985; KAMPS and SEFTON 1986), consistent with the central role of this residue in enzyme activity.

Several highly conserved regions within the catalytic domain have been implicated as key structural elements of the active site (Fig. 3). The sequence Asp-Phe-Gly (residues 404–406) is one of the most conserved structures among the protein kinases (HANKS et al. 1988). It has been reported that the equivalent of Asp-404 in cAMP-dependent protein kinase, Asp-184, may participate in the interaction with phosphate groups of ATP through Mg^{2+} salt bridges (BRENNER 1987). The sequence Ala-Pro-Glu (430–432) is often found conserved in serine/threonine and tyrosine kinases. Point mutations changing any one of these residues in pp60$^{v\text{-}src}$ (Table 3) block transformation and severely reduce kinase activity (BRYANT and PARSONS 1984), suggesting that this region plays an important role in maintaining structure of the catalytic domain and may reside near the catalytic site. Two additional sequence motifs specifically conserved in either the serine/threonine or the tyrosine kinases are thought to play a role in the recognition of the correct hydroxyamino acid. The sequence Asp-Leu-Arg(or Ala)-Ala-Ala-Asn (residues 386–391) is conserved in all vertebrate members of the tyrosine kinase family and differs from a consensus sequence found in most serine/threonine kinases (HANKS et al. 1988). Similarly, another such sequence (residues 425–432) includes the conserved Ala-Pro-Glu motif and is conserved among tyrosine kinases but not among the serine/threonine kinases (HANKS et al. 1988). Finally, the site of pp60^{src} autophosphorylation, Tyr-416, is conserved in all tyrosine protein kinases and is presumed to play an important role in tyrosine kinase structure and function. Surprisingly, mutation of Tyr-416 in pp60$^{v\text{-}src}$ does not abolish in vitro transforming ability (CROSS and HANAFUSA 1983; SNYDER et al. 1983), although such mutants are compromised in their ability to induce tumors in animals (SNYDER and BISHOP 1984). KMIECIK et al. (1988) have reported that mutation of Tyr-416 to Phe in pp60$^{v\text{-}src}$ reduces the in vitro kinase activity fivefold, and suggest that in spite of the reduction of kinase activity, the high level of expression of the less-active protein is still sufficient to induce the transformation of cells in culture, but perhaps not under the more stringent conditions in vivo.

In general, insertional mutagenesis of the catalytic domains of tyrosine kinases (STONE et al. 1984; REES-JONES and GOFF 1988) including *src* (DECLUE and MARTIN

1989), has confirmed the view that much of the domain is required for efficient enzyme activity. There are, however, some interesting examples in which insertions do not inactivate transforming function. For example, DECLUE and MARTIN (1989) have shown that a linker insertion at position 356 does not inhibit *src* transformation of chicken cells in culture (Table 3). This mutation resides close to a region deleted in the ts mutant tsNY 68 (residues 355–358) (NISHIZAWA et al. 1985; MAYER et al. 1986). This observation would argue for a certain structural plasticity within this region.

Biologically isolated mutants in the kinase domain have in general been temperature-conditional for kinase activity, and map within the "nonconserved" sequences in the kinase domain. The mutants isolated from the background of the SR strain of RSV display a consitent pattern: they all contain a Val to Met change at position 461, as well as one or more mutations at other sites within the kinase domain (MAYER et al. 1986) (Table 3). As mentioned above, the 461 mutation is phenotypically neutral by itself, but appears to potentiate the effects of mutations at other sites. For example, tsNY 68 contains the 461 mutation and a deletion of amino acids 351–354. Chimera construction demonstrates that neither mutation is sufficient for temperature sensitivity of transformation or kinase activity (NISHIZAWA et al. 1985; MAYER et al. 1986). tsPA 104 contains the 461 mutation and a Leu to Pro change at residue 325; tsNY 72 has the 461 mutation and a Pro to Ser at 503 (MAYER et al. 1986); tsCU 11 has the 461 mutation and a Thr to Ile at 508 as well as a Gly to Ala at 236 (D. WAGES, J. KEEFER, A. ESPESETH, T. PARSONS, and M. WEBER, unpublished data). Thus, SR strain temperature-conditional mutants isolated independently in three different laboratories all display this common mutation at position 461.

Interestingly, two partially transforming mutants which are not temperature sensitive carry the same 461 mutation: CU 2 and CU 12 (Table 3). As mentioned above, CU 12 is a fusiform mutant with a second-site mutation within the A region. CU 2 carries two other mutations within the kinase domain, at positions 403 and 413. Cells infected with CU 2 display elevated levels of cellular tyrosine protein phosphorylation, but are specifically defective in phosphorylation of calpactin (NAKAMURA and WEBER 1982; COOPER et al. 1983a). Thus, the pp60^{src} encoded by this mutant appears altered in either its accessibility to or its recognition of a specific cellular substrate.

The temperature-conditional kinase mutants isolated in the Pr strain of RSV and sequenced by WYKE and his colleagues have yielded a somewhat more complex pattern. The critical mutation for temperature sensitivity in tsLA 29 resides at position 507 (although mutations at other sites potentiate this effect) (WELHAM and WYKE 1988). A second site mutation (Lys to Arg) at position 427 reverses temperature sensitivity, so that *src* behaves like wild type (WELHAM and WYKE 1988). TsLA 24 contains an Arg to His mutation at position 480, which can be reverted by a His to Tyr at 492. TsLA 31 carries a mutation at position 478 (Gly to Asp) which is reverted to wild type by mutations either at 492 (the same change seen with the LA 24 revertant) or at 377. Thus, analysis of naturally occurring kinase domain mutants suggests that the nonconserved portions of the carboxy-terminal half of the kinase domain (in particular the region around Val-461 in SR RSV) may play an important, but subtle, role in controlling the overall structure and thus the

Table 4. Mutations activating the oncogenic potential of pp60$^{c\text{-}src}$

Mutant[a]	Sequence alteration	Ref.	Morphology[b]	Anchorage-independent growth	Kinase Activity[c]		mT antigen binding	Comments
					In vitro	In vivo		
	c-src	1	normal	—[d]	—	—		
NY BB4	*v-src*432/*c-src*		T	+	A^+E^+	+		Transformation measured by infection of chicken embryo cells. All activating mutations elevate in vitro and in vivo pp60^{src} kinase activity
NY 701	*c-src*432/*v-src*		T	+	A^+E^+	+		
NY 801	*v-src*258/*c-src*		T	+	A^+E^+	+		
NY HB5	*v-src*79/*c-src*		td	—	—	+		
NY 851	*v-src*110/*c-src*		T	+	A^+E^+	+		
NY 901	*v-src*179/*c-src*258 *v-src*432/*c-src*		T	+	A^+E^+	+		
NY 951	*c-src*258/*v-src*432/*c-src*		T	+	A^+E^+	+		
NY 378	E378 → G	2	T	+	A^+E^+	+		Isolated as spontaneous transforming mutants of *c-src*
NY 501T7-M	I441 → F		T	+	A^+E^+	+		
P*src* 527	Y527 → F	3	T	+	E^+	N.R.		Transformation measured by focus formation on NIH3T3 cells. Kinase activity of 416, 527F double mutant slightly greater than *c-src*
P*src* 416/527	Y416, 527 → F		normal	—	E^+	N.R.		
F 527	Y 527 → F	4	T	+	E^+	N.R.	+	Transformation determined by focus formation on NIH3T3 cells. Kinase activity of Am517 similar to that of *c-src*
Am 517	E517 → TER		T	+	—	N.R.	—	
LJ 527	Y527 → F	5	T	+	E^+	+		Transformation determined by NIH3T3 focus assay. Kinase activity of 416, 527F double mutant measured relative to 527F
LJ 519/527	Y519, 527 → F		T	+	E^+	+		
LJ 416/527	Y416, 527 → F		normal	—	+/—	—		

CH 527F	Y527 → F	6	T	+	NR	+	+	Transformation determined by infection of chicken embryo cells
CH 518 Am	D518 → Am		T	+	NR	+	—	
CH 523 Am	T523 → Am		T	+	NR	+	NR	
CH 530 Am	G530 → Am		normal	—	NR	—	+	
CH 90F	Y90 → F	7	slight	+/—	increased	increased		Transformation assayed by infection of chicken embryo cells. Cells infected with this group of mutants exhibited an intermediate transformed phenotype, characterized by an altered morphology, increased (but not wt) growth in soft agar, and increased in vivo and in vitro kinase activity
CH 92F	Y92 → F		slight	+/—	increased	increased		
CH 95W	R95 → W		slight	+/—	increased	increased		
CH dl92	*dl*92–95		f	+/—	increased	increased		
SB NX	*dl*15–89	8	spindle	+	increased	increased		Mutants NP and BS showed a restricted pattern of tyrosine phosphorylation of cellular proteins
SB NP	*dl*15–225		f	+	increased	restricted		
SB HP	*dl*112–225		normal	—	—	—		
SB BS	*dl*55–169		f	+	increased	restricted		

[a] See Table 1 for definition of symbols; boundaries of chimeric *src* proteins are designated by amino acid numbers. *am*, amber termination codons.
[b] Cells overexpressing avian c-*src* exhibit no appreciable alteration and hence have been termed normal.
[c] In vitro kinase activity measured relative to overexpressed c-*src*
[d] c-*src*-infected cells may form small restricted colonies under certain conditions.

References: 1, Kato et al. 1986; 2, Levy et al. 1986; 3, Kmiecik and Shalloway 1987; 4, Cartwright et al. 1987a; 5, Piwnica-Worms et al. 1987; 6, Reynolds et al. 1987; 7, Potts et al. 1988; 8, Nemeth et al. 1989.

heat sensitivity of the kinase domain. The precise influence of individual mutations upon the tertiary structure of pp60^{src} will have to await detailed X-ray crystallographic analysis of the protein.

The carboxy-terminus of the catalytic domain has been positioned by deletion analysis of the v-*src* gene at or near Leu-516 (WILKERSON et al. 1985; YACIUK and SHALLOWAY 1986). Frameshift mutations at residue 503 or 515 yield transformation-defective variants with reduced in vivo and in vitro kinase activity (WILKERSON et al. 1985). Insertion or deletion mutations at amino acid 516 also block transformation and inhibit tyrosine kinase activity (YACIUK and SHALLOWAY 1986). As will be discussed below, this residue also marks the site of the recombinational event giving rise to the novel carboxy-terminal sequence of viral *src* isolates (Fig. 1).

5 Genetic Analysis of Cellular *src* Protein

5.1 Carboxy-Terminal Regulatory Domain

The carboxy terminus of c-*src* contains 19 amino acids missing from the viral *src* gene. Present evidence shows that these amino acids comprise a regulatory domain responsible for the negative regulation of pp60^{src} activity. The major site of in vivo tyrosine phosphorylation of pp60$^{c\text{-}src}$ is a carboxy-terminal Tyr residue, Tyr-527 (COOPER et al. 1986; LAUDANO and BUCHANAN 1986). Evidence for the negative regulation of pp60$^{c\text{-}src}$ activity was first presented by COURTNEIDGE (1985), who showed that pp60$^{c\text{-}src}$ tyrosine kinase activity is inhibited by tyrosine phosphorylation. Further evidence for regulation was provided by COOPER and KING (1986), who reported that phosphatase treatment of pp60$^{c\text{-}src}$ leads to the dephosphorylation of Tyr-527 and activation of tyrosine protein kinase activity. The analysis of pp60$^{c\text{-}src}$ in polyoma virus-transformed cells (BOLEN et al. 1984; CARTWRIGHT et al. 1985) further demonstrated that an activated form of pp60$^{c\text{-}src}$ found associated with polyoma middle T (mT) antigen contains no detectable phosphorylation on Tyr-527, whereas free pp60$^{c\text{-}src}$, not associated with mT antigen, is phosphorylated on Tyr-527. These experiments taken together provide strong evidence that phosphorylation of Tyr-527 contributes to the regulation of pp60$^{c\text{-}src}$ activity in vivo and in vitro.

The use of site-directed mutagenesis to alter the structure and environment of Tyr-527 has provided additional evidence for the regulatory role of the carboxy-terminal domain in c-*src* function (Table 4). The alteration of Tyr-527 to Phe or Ser activates in vivo tyrosine kinase activity of pp60$^{c\text{-}src}$ and induces transformation of chicken or rodent cells in culture (CARTWRIGHT et al. 1987a; KMIECIK and SHALLOWAY 1987; PIWNICA-WORMS et al. 1987; REYNOLDS et al. 1987). In contrast, other mutations within the carboxy-terminus, e.g., Tyr-511 to Ser or Tyr-519 to Phe, do not induce transformation or activate tyrosine kinase activity. The introduction of *amber* mutations at positions 517, 518 or 523 which alter the carboxy-terminus by causing premature termination and truncation of the c-*src* protein also activate tyrosine kinase and transforming activities (REYNOLDS et al. 1987; YACIUK et al. 1988), whereas an *amber* mutation at position 530 has little effect on c-*src* function. Even though the alteration of Tyr-527 to Phe activates transformation of NIH 3T3 cells, the introduction of a second mutation at Tyr-416 (to Phe) suppresses the activation of transforming activity (KMIECIK and SHALLOWAY 1987; PIWNICA-WORMS et al. 1987). These experiments clearly demonstrate that the alteration of carboxy-terminal structure by the mutation of Tyr to Phe is sufficient to activate both transforming and tyrosine kinase activity. However, the fact that mutations at Tyr-416 suppress this activation strongly argues that the autophosphorylation activity of pp60$^{c\text{-}src}$ is important in mediating activities (phosphorylation of substrates?) leading to transformation. These observations are in contrast to mutagenesis experiments with pp60$^{v\text{-}src}$ in which mutation or deletion of Tyr-416 did not inhibit focus formation, but led to alterations in the efficiency of tumor formation (CROSS and HANAFUSA 1983; SNYDER et al. 1983; SNYDER and BISHOP 1984). It is likely that other mutations present in the v-*src* gene play a role in enhancing the transforming activity of its gene product, rendering unnecessary the possible structural changes induced by autophosphorylation.

Additional studies have attempted to delineate the minimal sequences required to direct the efficient phosphorylation of Tyr-527. CHENG et al. (1988) showed that mutation of residues adjacent to Tyr-527 (Table 4) does not activate transformation or lead to the decreased phosphorylation of Tyr-527, indicating that these sequences play only a minor role in the recognition by the kinases and phosphatases. However, three mutants which have in common the modification of Glu-524 are phosphorylated on both Tyr-416 and Tyr-527 and are weakly transforming in NIH 3T3 cells, suggesting that other mechanisms in addition to complete dephosphorylation of Tyr-527 can lead to activation of pp60$^{c\text{-}src}$. The requirement for correct orientation of the carboxy-terminus relative to the kinase domain has been investigated by introduction of "spacer" mutations that change the distance between Tyr-527 and the carboxy-terminus of the kinase domain (B. COBB, A. REYNOLDS, M. PAYNE, and T. PARSONS, unpublished results). Either the deletion or insertion of peptide sequences at amino acid position 518 efficiently activates the transforming activity of pp60$^{c\text{-}src}$. In addition, phosphorylation of Tyr-527 was observed in several of these activated mutants, containing both insertions or deletions, suggesting that the tertiary structure of the carboxy-terminus is required for the negative regulation of c-*src* tyrosine kinase activity and that the juxtaposition of Tyr-527 with the active site of pp60^{src} may be an important factor in the regulation of activity. MACAULEY and COOPER (1988) have recently shown that the carboxy-terminus of p56 *lck* can regulate the activity of pp60^{src}, even though there are some five amino acid differences between the carboxy-termini of *lck* and *src* proteins. These results further support the notion that regulation of kinase activity requires the proper three-dimensional folding of carboxy-terminal residues, a feature of the structure that may be influenced significantly by phosphorylation/dephosphorylation of Tyr-527.

5.2 Activating Mutations in Other Domains of *src*

The role of other structural domains in the regulation of c-*src* activity has been investigated by analyzing spontaneous transformation-competent revertants of c-*src*, construction of v-*src* and c-*src* chimeric genes, and by the site-directed mutation of specific c-*src* sequences. Continued passage of viruses containing the nononcogenic c-*src* gene in chicken cells gives rise to low titers of transforming virus. Cells infected with these "revertant" transforming viruses exhibit a fully transformed morphology and display higher levels of tyrosine kinase compared to the parental c-*src* infected cells (Levy et al. 1986) (Table 4). Molecular cloning and sequencing of two such isolates (Levy et al. 1986) revealed point mutations resulting in single amino acid changes within the kinase domain of each "activated" *src* protein. One isolate (NY 378) contained a Glu to Gly change at position 378; the second isolate (NY 501) contained an Ile to Phe at position 441. These results suggest that small alterations within the catalytic domain can activate the kinase activity and subsequent transforming activity of the c-*src* protein. In light of the fact that the carboxy-terminal regulatory domain appears functionally to suppress the active site of the c-*src* protein, it is interesting to speculate that such alterations in the catalytic domain may serve to perturb the interaction of the carboxy-terminal regulatory domain with the catalytic domain, giving rise to unregulated $pp60^{c\text{-}src}$.

Sequence alterations within the modulatory domain of $pp60^{c\text{-}src}$ also activate the transforming activity of the c-*src* gene product. A comparison of the sequence differences between c-*src* and the different strains of v-*src* reveals a common sequence alteration at residue 95 (Arg to Trp) in the SR, Pr, and B77 RSV. The Br strain, although it lacks the Arg to Trp alteration, contains changes at positions 91 and 96. The possible significance of these alterations was first indicated using chimeric *src* genes engineered by exhanging restriction fragments from c-*src* and v-*src*. Cells infected with *src* variants encoding *src* proteins containing the first 110 (but not the first 84) amino acids of v-*src* and the remaining amino acids from c-*src* or the single Thr to Ile change at position 338 exhibited an altered morphology, as well as increased grwoth in soft agar (Kato et al. 1986) (Table 4). In vitro mutagenesis of amino acids 90–95 showed that the alteration of amino acid residues 90 (Tyr to Phe), 92 (Tyr to Phe), 95 (Arg to Glu, Lys Trp, Gly), or a deletion of residues 92–95 induced a transformed phenotype (Potts et al. 1988). Surprisingly, the mutation of Arg 95 to Gln had no effect on morphology. Characterization of cells infected with the transforming c-*src* variants revealed that a variety of parameters of transformation were affected by these mutations, including glucose transport, growth in agar, and in vivo tyrosine phosphorylation. In all cases, however, the mutant infected cells exhibited only intermediate activation of the individual transformation parameter. Similar mutations have been isolated in this region by L. Fox and J. S. Brugge (personal communication) and exhibit similar phenotypes. The observation that sequence alterations within the A region of the homology domain of c-*src* partially alter growth and cell morphology suggests that this region may directly influence the tyrosine kinase activity of the *src* protein by modulation of tertiary structure or alter the interaction of *src* with another protein or proteins necessary for the regulation of kinase activity.

Nemeth et al. (1989) have used deletional mutagenesis of c-*src* to further investigate the role of the modulatory region in $pp60^{c\text{-}src}$ regulation and function. Cells expressing a c-*src* variant containing a deletion of residues 15–89 (a deletion of virtually all of the unique domain) displayed a refractile, spindle-shaped morphology, enhanced anchorage-independent growth, and elevated levels of phosphotyrosine-containing proteins. In contrast, deletion of amino acids 112–225 (most of the homology domain) did not activate the transforming ability of the *src* protein or increase tyrosine kinase activity. Other deletion variants, residues 15–225 or 55–169, exhibited a partially transformed phenotype characterized by a fusiform morphology, formation of diffuse colonies in soft agar, and increased tyrosine phosphorylation of only a few cellular proteins. These results are significant, in that deletion of the unique domain of c-*src* leads to significant activation of transforming potential. The results also suggest that the c-*src* gene product displays an interactive domain structure similar to that of the v-*src* gene product. Clearly, additional mutagenesis studies are required to assess the contributions of each domain to the regulation of c-*src* activity.

5.3 Middle T Antigen Binding

The binding of mT antigen to $pp60^{c\text{-}src}$ and the activation of *src* kinase activity results from the interaction of mT antigen with the carboxy-terminal regulatory domain of $pp60^{src}$. Either the premature termination of the c-*src* protein at residue 516 or 519 (Cartwright et al. 1987a; Cheng et al. 1988), or the substitution of carboxy-terminal c-*src* sequences with carboxy-terminal sequences of v-*src*, blocks the interaction of mT-antigen with *src* protein (Cheng et al. 1988). Analysis of point mutations within the carboxy-terminus of c-*src* indicates that the amino acid sequences immediately surrounding Tyr-527 are not required for mT antigen binding. The analysis of truncated forms of $pp60^{c\text{-}src}$ (Table 4), however, reveals that a region bounded by residues Asp-518 and Pro-525 is required for efficient mT antigen binding (Cheng et al. 1988). The proximity of the mT antigen binding site to the site of regulatory Tyr-527 phosphorylation suggest that mT antigen may activate $pp60^{c\text{-}src}$ by masking the phosphorylation of Tyr-527. Alternatively, mT antigen binding may alter the conformation of the carboxy-terminal domain, thereby activating $pp60^{c\text{-}src}$ and promoting the accessibility of Tyr-527 to dephosphorylation. It is tempting to speculate that such a mechanism may be used by cellular components to regulate c-*src* activity, although attempts to date to identify such proteins have not thus far been successful.

5.4 Tyrosine 527 Phosphorylation: Autophosphorylation or a Regulatory Kinase?

The fundamental importance of Tyr-527 phosphorylation in the regulation of $pp60^{c\text{-}src}$ activity has led to experiments designed to test whether $pp60^{c\text{-}src}$ itself or another tyrosine protein kinase phosphorylates Tyr-527. Jove et al. (1987) showed that an ATP binding site mutant constructed by alteration of Lys-295 to Met lacks detectable kinase activity, whether expressed in yeast cells or chicken cells. Upon expression of the mutant protein in chicken cells, Tyr-527 is phosphorylated to levels similar to unmutated $pp60^{c\text{-}src}$, leading the authors to conclude that chicken cells contain a protein kinase capable of phosphorylating $pp60^{c\text{-}src}$ *in trans*. The expression of mutant protein in yeast cells, however, does not yield significant Tyr-527 phosphorylation of the c-*src* protein, suggesting that such a protein kinase is absent in yeast (Jove et al. 1987). Expression in yeast cells of two mutant *src* genes, each lacking a site of tyrosine phosphorylation (Tyr-416 or Tyr-527), either together or separately, showed that pp60$^{c\text{-}src}$ molecules could phosphorylate each other at Tyr-416 and Tyr-527 (Cooper and MacAuley 1988). Experiments prohibiting the intramolecular phosphorylation suggested that intermolecular phosphorylation of Tyr-416 and Tyr-527 predominates in yeast cells lacking a regulatory kinase. Similarly, pp60$^{c\text{-}src}$ expressed in a baculovirus system is phosphorylated to a small extent (less than 10% of the molecules), lending further support to the inefficiency of autophosphorylation (H. Piwnica-Worms, J. Kaplan and David Morgan, personal communication). As of this writing the elusive regulatory kinase remains unidentified, although the availability of large amounts of purified pp60$^{c\text{-}src}$ generated by the insect-vector system will undoubtedly speed in the search for such a kinase.

6 Genetic Analysis of Transformation by *src*: The Search for Substrates

A major goal in studies of *src* genetics is to increase understanding of the molecular mechanisms by which expression of *src* alters cellular phenotypes. Because the tyrosine kinase activity of pp60^{src} is necessary for its transforming activity, it is widely assumed that tyrosine phosphorylation of cellular substrate proteins plays an essential role in transformation. Accordingly, in order to understand oncogenesis by *src* and other tyrosine kinase oncogenes, it will be necessary to identify and characterize biologically significant substrates for pp60^{src}. Potentially important tools in the identification of such significant substrates are pp60^{src} mutants defective in their ability to recognize and phosphorylate specific cellular proteins while retaining high levels of intrinsic protein kinase activity, i.e., kinase-positive/transformation-defective mutants. Such mutants are presumed to be defective in the phosphorylation of one or more essential cellular target proteins, while still retaining the ability to phosphorylate many adventitious substrates. In this section we will briefly review the known kinase-positive/transformation-defective mutants, with an emphasis on identifying regions in the *src* gene which are critical for recognition of cellular target proteins. We then will outline the current state of knowledge concerning the best-characterized candidate substrates for pp60$^{v\text{-}src}$ and the use of kinase-positive/transformation-defective *src* mutants to characterize the biological significance of these tyrosine phosphorylations.

6.1 Substrate Recognition by pp60$^{v\text{-}src}$

Among the first kinase-positive/transformation-defective *src* mutants to be identified were biologically isolated mutants which induced the appearance of some parameters of transformation better than others (Anderson et al. 1981; Becker et al. 1977; Calothy et al. 1978; Weber and Friis 1979; Weber 1984). They could be distinguished from quantitatively "leaky" mutants in two ways: First, the pattern of phenotypes induced by each mutant could not be mimicked simply by varying quantitatively the amount of pp60^{src} kinase activity (e.g., by placing cells infected with thermosensitive kinase mutants at various temperatures intermediate between the restrictive and permissive temperatures). Second, given a panel of such mutants, the patterns of phenotypes they induced could not be ordered into a "hierarchy" of degrees of transformation. The biological properties of these *src* mutants indicate that not only are they altered in substrate selection, but that several cellular proteins are biologically significant targets for pp60$^{v\text{-}src}$, different targets being involved in generating different manifestations of the transformed state (WEBER 1984). *src* mutants of this type were originally termed "partially transforming" mutants (BECKER et al. 1977), but it might be more accurate to refer to them as "phenotype-dissociation" mutants, since they disscociate the induction of one phenotype from another.

Fusiform mutants are among the best characterized of these "phenotype dissociation" mutants. While inducing many manifestations of transformation (e.g., anchorage-independent growth), in all the cases which have been examined they fail to induce the loss of surface fibronectin (ROHRSCHNEIDER and REYNOLDS 1985), even though this is a transformation parameter which is exquisitely sensitive to low levels of *src* activity (determined using thermosensitive kinase mutants) (WEBER and FRIIS 1979). Presumably, the biological properties of the fusiform mutants reflect a deficiency in the ability of the mutant pp60$^{v\text{-}src}$ to interact with a specific subset of cellular proteins, which directly or indirectly regulate fibronectin gene expression (ROHRSCHNEIDER and REYNOLDS 1985)). Interestingly, fusiform mutants are defective, not only at extinguishing fibronectin gene expression, but also at inducing expression of a mitogenic platelet protein gene (SUGANO et al. 1987). Most of the fusiform mutants map to the modulatory region of *src*, particularly to the A and B domains in the homology region. Thus, this region apparently plays a role in determining the specificity of pp60$^{v\text{-}src}$.

The modulatory region is also the region in which host-range mutants of *src* most often appear. These mutants are able to transform chicken cells but not rodent cells. Presumably they are altered in the ability to interact with a cellular protein which differs subtly between the avian and the mammalian cells.

Mutations in the B domain of the modulatory region also can drastically affect the rate at which pp60$^{v\text{-}src}$ turns over (WANG and PARSONS 1989). This may be due to changes in the interaction of the *src* protein with a cellular proteolytic system, or it may be due to changes in the interaction of the mutant pp60$^{v\text{-}src}$ with other cellular proteins which stabilize or protect it. In either case, it points to the importance of this region of *src* for interactions with cellular proteins.

Thus, three sorts of *src* mutants thought to display phenotypes arising from altered interactions with cellular proteins map to the modulatory region: fusiform mutants, host-range mutants and turnover mutants. The fact that many of these mutant proteins retain a near-normal kinase-specific activity supports the suggestion that they are altered in some way in target specificity. Whether the relevant targets are in fact substrates for tyrosine phosphorylation, or whether they are proteins with which pp60$^{v\text{-}src}$ interacts but does not phosphorylate, remains to be seen. It is noteworthy that most of these mutants fall within the region of *src* which shows homology with PLC-148 and the *crk* oncogene, neither of which is a kinase.

Although it seems highly probable that the "modulatory" region is involved in cellular interactions, the properties of at least one *src* mutant indicate that the kinase domain also plays a role in determining specificity. The "phenotype-dissociation" mutant CU 2 induces a pattern of phenotypes opposite that induced by the fusiform mutant CU 12: cells infected with CU 2 have lost their surface fibronectin, but still require anchorage for growth. This mutant is specifically defective in phosphorylation of calpactin I in vivo, yet it retains 50% of the wild-type level of overall kinase activity, and thus is altered in substrate specificity. CU 2 contains three mutations, all within the kinase domain (Table 3).

Mutations in *src* could affect substrate selection directly, by affecting the binding of pp60$^{v\text{-}src}$ to cellular substrate proteins, or indirectly, by altering the location of pp60$^{v\text{-}src}$ in cells. The latter category of mutants is clearly exemplified by the unmyristylated mutants, which render pp60^{src} cytoplasmic, affecting both the transforming capacity and (as described below) the pattern of substrate phosphorylation (Kamps et al. 1985, 1986; Linder and Burr 1988b; Hamaguchi et al. 1988; Reynolds et al. 1989). Interestingly, at least two of the phenotype-dissociation mutants, CU 2 and CU 12, also display alterations in intracellular distribution (Rohrschneider and Rosok 1983). The CU 2 *src* protein is highly concentrated in adhesion plaques, whereas the CU 12 protein is absent from adhesion plaques. Although it is not known whether these unusual intracellular localizations are a cause or a consequence of the unusual cellular morphologies induced by these mutants, it is quite possible that the ability of the mutant proteins to phosphorylate specific targets is affected by their cellular localization.

The overall picture of pp60$^{v\text{-}src}$ which emerges from this sort of analysis is of a protein with strongly interacting domains, all of which can influence interaction with specific cellular proteins. The modulatory region seems very likely to be critical for recognition of cellular proteins, but mutations in this region can also affect kinase activity. The kinase domain, though essential for enzymatic activity, also affects localization of pp60$^{v\text{-}src}$ and substrate specificity; the myristylation signal sequence determines the localization of the protein, thereby affecting its pattern of phosphorylations and, hence, its function. These varied kinase-positive/transformation-defective *src* mutants, occurring in various regions of the gene and having various biological effects on cells, are potentially very useful — particularly when used as a panel — for correlating specific phosphorylations with transformation phenotypes, and thus distinguishing adventitious from significant phosphorylations.

6.2 From Phosphorylation to Phenotype

Identification of functionally significant substrates for the tyrosine kinase activity of pp60^{src} has been a difficult task, because tyrosine phosphorylation in *src*-transformed cells is both promiscuous and nonabundant: Nonabundant because no more than about 1% of the total phosphoamino acid in *src*-transformed cells is on tyrosine, and promiscuous because many of the tyrosine phosphorylations detected appear to be adventitious. In addition, activation of other tyrosine kinase by pp60$^{v\text{-}src}$ could contribute to the overall complexity of observed tyrosine phosphorylated proteins.

Until recently, most investigations of tyrosine phosphorylation relied on labeling cells with $^{32}P_i$, followed by either phosphoamino acid analysis (to separate phosphotyrosine from phosphothreonine and phosphoserine) (Martinez et al. 1982;

BEEMON et al. 1982) or treatment of labeled gels with KOH (to remove RNA and preferentially hydrolyze phosphoserine and phosphothreonine) (COOPER and HUNTER 1981). More recently, phosphotyrosine-specific antibodies of high titer and specificity have been developed, which provide a sensitive and effective tool for identifying and isolating tyrosine-phosphorylated proteins (COMOGLIO et al. 1984; KAMPS and SEFTON 1988). These antibodies have greatly diminished the problem of detecting tyrosine-phosphorylated proteins. However, quantitative and occasional qualitative differences are apparent in the tyrosine phosphorylation pattern detected with phosphotyrosine-specific antibodies compared to that obtained using ^{32}P-labeling followed by either KOH treatment or direct amino acid analysis. For example, Western blots with phosphotyrosine-specific antibodies display predominant phosphorylations in the mass region greater than 100 kDa (KAMPS and SEFTON 1988; REYNOLDS et al. 1989; our unpublished results), whereas amino acid analysis of slices of gels of electrophoresed cell proteins indicates that the majority of the tyrosine phosphorylations are on proteins of between 30 and 100 kDa (MARTINEZ et al. 1982). It is unclear to what extent this quantitative difference is a consequence of inefficient elution of high-M_r proteins from gel slices (prior to amino acid analysis), and to what extent it is due to the biased specificity of phosphotyrosine-specific antisera. In any event, it is important to recognize that no single detection technique provides a complete, accurate, and quantitative picture of the pattern of in vivo tyrosine phosphorylations.

Utilization of either phosphotyrosine-specific antibodies or in vivo labeling with $^{32}P_i$ reveals a very large number of tyrosine-phosphorylated proteins in *src*-transformed cells: at least 50 are readily detected. In order to understand the biochemical mechanisms involved in transformation by *src*, it will be necessary to determine which of these many phosphorylations are biologically significant and which are adventitious. The use of kinase-positive/transformation-defective *src* mutants has provided a substantial, although still very incomplete, body of information which can be used to help achieve that goal. By comparing the pattern of tyrosine phosphorylations induced by such mutants with that obtained using transformation-competent *src*, one can hope to identify proteins whose phosphorylation correlates with transformation. The correlation is greatly strengthened if it is seen in cells infected with various kinase-positive/transformation-defective mutants of differeing molecular nature. Although correlations cannot be taken as proof of causality, the failure of a correlation can suggest that a particular protein phosphorylation is insufficient for transformation, unnecessary for transformation, or both (COOPER et al. 1983a).

We will outline below current knowledge concerning the proteins known to become tyrosine-phosphorylated in *src*-transformed cells, emphasizing, where available, the information provided by *src* genetics concerning the functional significance of those phosphorylations.

6.2.1 Calcium Binding Proteins

Calpactin I. RADKE and MARTIN (1979) were the first to identify a change in cellular protein phosphorylation which accompanied transformation by *src*: a 36 kDa protein identified by two-dimensional gel electrophoresis became phosphorylated when *src* was expressed. It subsequently was determined that the phosphorylation was on tyrosine (as well as serine), and that approximately 10% of the protein becomes phosphorylated (RADKE et al. 1980). Cloning and sequencing the gene for this 36-kDa protein revealed that it is a member of a family of calcium- and lipid-binding proteins, which includes a 35-kDa substrate for the EGF receptor (reviewed in BRUGGE 1986b, CROMPTON et al. 1988). The candidate pp60$^{v\text{-}src}$ substrate was termed calpactin I.

Calpactin I is a tetrameric protein, consisting of two 36-kDa subunits (heavy chains) and two 10-kDa subunits, which have strong homology to the brain calcium-binding protein, S-100. Phosphorylation of the calpactin I heavy chain occurs in an amino-terminal region of the protein which is near the site of phosphorylation by PKC (GOULD et al. 1986b). Interestingly, examination of isoelectric shifts indicates that any given molecule of calpactin is phosphoralyted on only one site (RADKE et al. 1980), as if phosphorylation on tyrosine and serine were mutually exclusive. The fact that *src* expression leads to phosphorylation on serine as well as tyrosine suggests that the tyrosine phosphorylation on one heavy chain renders the other heavy chain in the tetramer susceptible to phosphorylation by another kinase. Neither the functional significance of these phosphorylations nor the nature of the proximal kinases is known with certainty.

Calpactin I is an abundant protein (approximately 0.1% of the total cellular protein in fibroblasts) and is localized in a submembranous "cortical" region of the cell. It copurifies with membranes, but also is found in the cytoskeleton-enriched, detergent-insoluble fraction of cells — as is pp60$^{v\text{-}src}$ (CHENG and CHEN 1981; COURTNEIDGE et al. 1982; GREENBERG and EDELMAN 1983; KAJI and AMINI 1983; RADKE et al. 1983). The biochemical and physiological functions of this protein family are the subject of considerable current investigation: they may play a role in regulating phospholipid metabolism or in membrane fusion or interactions between membranes and the cytoskeleton.

The correlation of calpactin I phosphorylation with transformation has been studied in detail, using a wide variety of *src* mutants. NAKAMURA and WEBER (1982) and COOPER et al. (1983a) found that calpactin I phosphorylation was strikingly deficient in cells infected with the "phenotype-dissociation" mutant CU 2. Cells infected with this mutant fail to grow in soft agar and are poorly tumorigenic, but they have lost surface fibronectin and density-dependent growth control (ANDERSON et al. 1981; KAHN et al. 1982). By contrast to the CU 2-infected cells, cells infected with nonmyristylated mutants continued to display substantial levels of calpactin I phosphorylation, indicating that this phosphorylation is not sufficient for morphological transformation, anchorage-independent growth, or loss of fibronectin (KAMPS et al. 1986). Cells infected with the partially transforming mutant tsGI 251 display transformed growth properties at 42° and grow poorly at 35°, but show the reverse temperature dependence for calpactin phosphorylation (NAKAMURA and WEBER 1982; COOPER et al. 1983a). Thus, the extent of phosphorylation of calpactin I in *src*-transformed cells correlated very poorly with morphological changes, loss of fibronectin, and loss of density-dependent inhibition, being neither necessary nor sufficient for these changes. Its phosphorylation correlated best (although still imperfectly) with production of plasminogen activator and with tumorigenicity. But in spite of the disappointingly poor correlations between transformation and calpactin phosphorylation, the biochemical and physiological properties of this protein continue to make it an intringuing candidate for being a functional target of pp60^{g,src}. It is possible that phosphorylation of this protein plays a role in some aspect of transformation which has not been assayed, or that the quantitative relationship between calpactin phosphorylation and transformation is not linear, or that calpactin phosphorylation plays a contributory rather than a directly causal role in the genesis of the transformed state. However, the fact that calpactin and pp60$^{v\text{-}src}$ are both present in high concentrations in the same cellular compartments and the poor correlation between this phosphorylation and transformation make it at least as likely that this phosphorylation is adventitious.

Calmodulin. Phosphotyrosine has been found on calmodulin immunoprecipitated from transformed cell extracts (FUKAMI et al. 1986). The stoichiometry of this phosphorylation has not been reported, nor has its significance been investigated using *src* mutants. However, the central importance of calmodulin in regulating cellular responses to calcium indicates that this would be a worthwhile area for exploration.

6.2.2 Membrane Proteins

Because membrane association of pp60$^{v\text{-}src}$ is necessary for transformation, it is widely thought that an important target for pp60$^{v\text{-}src}$ would be a membrane protein. Accordingly, this is an area receiving a great deal of current attention.

Fibronectin Receptor. The fibronectin receptor is a member of a family of proteins termed "integrins" which link the extracellular matrix to the cytoskelton. Antibody against the fibronectin receptor precipitates a disulfide-linked complex of 160-, 140-, and 120-kDa glycoproteins, the lower mass forms of which display increased tyrosine phosphorylation in *src*-transformed cells (HIRST et al. 1986). Immunofluorescence also reveals an altered intracellular distribution of the receptor in the transformed cells. In vitro binding studies indicate that the interaction of the receptor with fibronectin or with talin (a cytoskeletal protein) is altered if the receptor is isolated from transformed cells (TAPLEY et al. 1989). However, it has not been shown that this functional change is due to phosphorylation. The stochiometry of phosphorylation of the receptor is not known, but appears to be very low, since the phosphorylation is not detected unless cells are incubated prior to lysis in orthovanadate, a potent inhibitor of phosphotyrosine phosphatases.

If phosphorylation of the fibronectin receptor was important for loss of surface fibronectin, one would expect this phosphorylation not to occur in cells which retain substantial quantities of surface fibronectin. Therefore, L. ROHRSCHNEIDER and collaborators (personal communication) examined the phosphorylation of the receptor in cells infected with CU 12, a partially transforming fusiform mutant which does not cause large losses of surface fibronectin. Phosphorylation of the receptor was indistinguishable from that seen in wild-type transformed cells, casting doubt on the functional relevance of this phosphorylation for fibronectin binding in vivo.

EGF Receptor. The EGF receptor becomes rapidly phosphorylated on tyrosine in *src*-transformed cells (WASILENKO and WEBER 1987). Production of transforming growth factor α (TGF-α) or similar factors which might stimulate that phosphorylation by an autocrine mechanism seems not to be involved. The stoichiometry of the phosphorylation is comparable to that obtained using saturating levels of EGF.

Variant cells which lack EGF receptors can still be transformed by *src* (Wasilenko and Weber 1987), indicating that the EGF receptor is not a necessary target for *src*-induced transformation. However, this receptor phosphorylation could be indicative of a class of membrane targets whose phosphorylation would induce alterations in growth, metabolism and morphology.

Unidentified Membrane Proteins. A 120-kDa protein which copurifies with plasma membranes is tyrosine-phosphorylated in wild-type transformed cells, but not in cells infected with nonmyristylayed mutants (Linder and Burr 1988a; Reynolds et al. 1989). The identity of this protein is uncertain, but it is not the fibronectin receptor or vinculin.

src-induced tyrosine phosphorylation has been detected on 95-kDa and 130-kDa proteins which are present in the glycoprotein fraction, isolated by lectin affinity chromatography (Monteagudo et al. 1984; L. Kozma and M. Weber, unpublished observations). These proteins do not become phosphorylated in cells infected with nonmyristylated mutants or in cells infected with the partially transforming mutant CU 2. The fact that phosphorylation of the 95- and 130-kDa glycoproteins correlates with transformation in cells infected with such diverse kinase-positive/transformation-negative mutants provides a strong impetus to identify their function.

6.2.3 Cytoskeletal Proteins

Because of the localization of pp60$^{v\text{-}src}$ in the cytoskeletal, detergent-insoluble fraction and the rapid effects of the protein on cellular morphology, it seems reasonable to suppose that some cytoskeletal proteins would become tyrosine-phosphorylated in *src*-transformed cells. Perhaps surprisingly, the major cytoskeletal proteins do not become detectably phosphorylated on tyrosine: actin, α-actinin, myosin, and tubulin (Sefton et al. 1981). The *src*-induced tyrosine phosphorylations occur on relatively less abundant proteins which might serve a role in regulating cell shape.

Vinculin. This 130-kDa protein is localized partially in adhesion plaques, as is some pp60$^{v\text{-}src}$. Approximately 1% of the vinculin in *src*-transformed cells becomes phosphorylated on tyrosine (Sefton et al. 1981), but whether the fraction which is in adhesion plaques displays a higher stoichiometry of phosphorylation is unknown. The relationship of vinculin phosphorylation to transformation has been examined in cells infected with a wide variety of *src* mutants by several laboratories (Antler et al. 1985; Kellie et al. 1986a, b; Rohrschneider and Reynolds 1985). In every case, no relationship between the degree of vinculin phosphorylation and any aspect of transformation was discernible.

Ezrin, Talin. Ezrin is a 81-kDa protein originally described as becoming tyrosine-phosphorylated in A431 cells treated with EGF (Gould et al. 1986). Talin is a 215-kDa protein which links integrins to the cytoskeleton (Pasquale et al. 1986; DeClue and Martin 1987). The possible role in transformation played by the phosphorylation of these proteins has not been explored, although the stoichiometry of their phosphorylation is low, and it is questionable whether low-level phosphorylation of a structural protein would be regulatory.

Unidentified Cytoskeletal Proteins. A 130-kDa protein and a group of proteins of approximately 65–70 kDa which are found in the detergent-insoluble, cytoskeleton-enriched fraction become tyrosine-phosphorylated in *src*-transfomed cells (Burr et al. 1981; Comoglio 1984; Gacon et al 1982). Their phosphorylation in this preparation is defective in cells infected with the fusiform *src* mutant, CU 12 (J. Burr, personal communication).

6.2.4 Proteins which Bind to pp60^{src}

Tyrosine-phosphorylated proteins which bind to pp60^{src} are good candidates for being direct substrates for this kinase.

130-kDa Protein. A 130-kDa tyrosine-phosphorylated protein has been detected in immunoprecipitates of pp60$^{v\text{-}src}$ from avian and rodent cells transformed by *src* (Lau 1986; Reynolds et al. 1989b). In RSV-infected chicken cells approximately 5%–10% of pp60^{src} is found in complex with p130. The stable association of p130 with pp60^{src} appears to be mediated through the homology domain since variant *src* proteins containing small deletions within this domain (e.g., CH *dl*165 and CH *dl*155) to not stably associate with p130 (Reynolds et al. 1989b). Unmyristylated pp60^{src} binds to p130 as well as to a smaller protein, p110. p130 is unrelated to the fibronectin receptor or to vinculin, and its relationship to the 130-kDa glycoprotein seen in lectin extracts is unknown. The functional significance of the complex between p130 and pp60^{src} and its possible role in transformation are under investigation.

50-kDa Protein. This protein binds to newly synthesized, soluble forms of pp60$^{v\text{-}src}$ (along with the p90 heat-shock protein), and is thus hypothesized to play a role in the transport of p60 to the membrane (Brugge 1986a). However, thermosensitive forms of pp60$^{v\text{-}src}$ display increased binding to p50 at the restrictive temperature, raising the possibility that the complex of p50, p60, and p90 is a scavenging system for inappropriately localized or partially denatured p60. Phosphorylation of p50 is increased in cells infected with thermosensitive kinase mutants and held at the restrictive temperature. This contradictory result is explained by supposing that the increased binding of p50 to pp60$^{v\text{-}src}$ at the restrictive temperature more than compensates for the diminished kinase activity at that temperature. In any event, phosphorylation of p50 is inversely related to transformation.

85-kDa Protein. A protein of 85 kDa coprecipitates with pp60$^{c\text{-}src}$ in a ternary complex with polyoma mT antigen (Courtneidge and Heber 1987; Kaplan et al. 1987). It has been hypothesized that this protein might be the phosphatidylinositol kinase which is found in that complex.

6.2.5 Cytosolic Proteins

Glycolytic Enzymes. Enolase, phosphoglycerate mutase, and lactic dehydrogenase become tyrosine-phosphorylated in *src*-transformed cells, and indeed the first two are among the more prominent tyrosine phosphorylations seen on two-dimensional gels (Cooper and Hunter 1981; Cooper et al. 1983b). However, the stoichiometry of phosphorylation of these abundant proteins is quite low (less than 1%), and none of these enzymes is rate-limiting for glycolysis. It thus is difficult to see how these phosphorylations could be regulatory. Not surprisingly, phosphorylation of phosphoglycerate mutase correlated very poorly with transformation in cells infected with a panel of *src* mutants (COOPER et al. 1983a). Although phosphorylation of enolase correlated moderately well with some aspects of transformation (the correlation was similar to that seen with calpactin I phosphorylation), there were several exceptions, and in this sort of analysis the exceptions are of more significance than the positive correlations. Moreover, enolase phosphorylation was increased in cells infected with nonmyristylated mutants (KAMPS et al. 1986). For these reasons we suspect that tyrosine phosphorylation of the glycolytic enzymes is adventitious.

42-kDa-Protein. Tyrosine phosphorylation of a cytosolic protein (or pair of proteins) of 42 kDa is commonly seen as a transient event in cells treated with a variety of mitogenic agents, including EGF, PDGF, insulin, insulin-like growth factor 2 (IGF-2), thrombin, or tetradecanoyl phorbol acetate (TPA) (COOPER et al. 1984b; NAKAMURA et al. 1983). This phosphorylation requires active PK C as an intermediary (VILA and WEBER 1988; KAZLAUSKAS and COOPER 1988). The same phosphorylation is often seen in cells transformed by *src* or other tyrosine kinase oncogenes (COOPER et al. 1981). Although a phosphorylation detected both in transformed cells and during normal mitogenesis is a good candidate for being involved functionally in growth control, the phosphorylation of this protein has not been seen in all laboratories or with all strains of RSV. It is possible that phosphorylation of this protein is important in establishing the transformed state, but that its phosphorylation is not maintained in long-term transformed cultures. In any event, the variability of its phosphorylation has made it difficult to determine its correlation with transformation.

6.2.6 Other Tyrosine-Phosphorylated Proteins

At least 50 phosphotyrosine-containing proteins can be identified in *src*-transformed cells using one-dimensional gel electrophoresis in combination with immunoblotting or immunoprecipitation with phosphotyrosine-specific antibodies (COMOGLIO et al. 1984; KAMPS and SEFTON 1988; LINDER and BURR 1988a, b; HAMAGUCHI et al. 1988; REYNOLDS et al. 1989). While it is apparent that these proteins include at least some of the tyrosine phosphorylations described above (e.g., calpactin) most are on proteins whose identity is unknown and whose role in transformation has yet to be explored. We expect that identification and characterization of these new phosphoproteins will provide interesting and novel insights into the mechanisms of transformation by $pp60^{v\text{-}src}$.

In summary, among the large number of tyrosine-phosphorylated proteins detected in *src*-transformed cells, several are proteins which display patterns of tyrosine phosphorylation which correlate well with transformation (e.g. pp 120, 95- and 130-kDa glycoproteins), and others have interesting and potentially regulatory properties (e.g., calpactin, calmodulin, mitogen, and matrix receptors). Some of the phosphorylations found on regulatory proteins may be less significant than they appear, since many of these proteins were examined precisely because of a bias that they might be involved in transformation. In either case, it is reasonable to continue to explore the relationship between the structure of *src*, the phosphorylation of these proteins, and the expression of transformation parameters in order to identify candidate target proteins and (more probably) to identify adventitious phosphorylations. Although these correlational studies can be helpful in reducing the number of protein phosphorylations under consideration, proof that any one of them is functionally involved in transformation will require the demonstration that phosphorylation alters the function of the protein in vivo and in vitro. Thus, the challenge we currently face is to isolate and molecularly to characterize these potentially important tyrosine-phosphorylated proteins, many of which are non-abundant and have unknown functions. Perhaps the combined use of phosphotyrosine-specific antibodies and purified pp60$^{v\text{-}src}$ will make possible the development of affinity-isolation reagents to assist with that task.

7 Concluding Remarks: Structure and Function of *src* Protein

In this review we have outlined the arguments for the domain structure of the *src* protein and have speculated about the functional significance of individual domains with respect to the process of cellular transformation. While the information presented supports the general concept of structural and functional domains, these data clearly point out the limitations of such an analysis in understanding, at the molecular level, the activities and regulation of pp60^{src}. The next problem is to understand how putative domains of the *src* protein interact with each other, as well as with cellular components to mediate events in transformation and in normal signal transduction. Solving the crystal structure of the *src* protein and defining the tertiary structure of "domains" will provide a unique opportunity to integrate genetic, biochemical and structural information.

Because mutations which inactivate pp60^{src} kinase activity also abolish transforming activity, it is generally assumed that phosphorylation of cellular proteins is necessary for transformation. We cannot conclude this review without pointing out the possibility that the kinase activity of pp60^{src} may serve only to autophosphorylate itself, a process that may be necessary in order to place pp60^{src} in an appropriate "activated/regulatory" conformation. The inability of kinase-negative mutants to transform would simply reflect the inability of pp60^{src} to achieve the "activated" conformation. It should be possible to test this hypothesis by selecting for or constructing putative kinase-negative/transformation-competent *src* mutants. If this hypothesis is correct, it would explain the difficulties to date of identifying biologically significant cellular substrates for pp60^{src}: there may not be any!

Finally, a number of very important questions remain to be addressed concerning the function of the normal *src* protein. Although substantive data exist to document the involvement of pp60$^{c\text{-}src}$ in the mitogenic response of cells to certain growth factors (Ralston and Bishop 1985; Gould and Hunter 1988; Luttrell et al. 1988; Wilson et al. 1989), the presence of relatively high levels of pp60$^{c\text{-}src}$ in neural cells and various secretory cells (Parsons and Creutz 1986; Golden et al. 1986; Ferrell and Martin 1988) poses provocative questions as to its function in these specialized cell types. In addition, other important questions remain as to the role of the *src* protein in cell differentiation (Alema et al. 1985), cell-cell communication (Azarnia et al. 1988), and even cell-cycle regulation (Chackalaparampil and Shalloway 1988). The fact that the *src* protein and other members of the *src* family of protein tyrosine kinases can be transduced by various animal retroviruses suggests that mutated *src* proteins may function as oncogenes in spontaneously arising malignancies. Indeed, increased levels of $pp60^{src}$ tyrosine kinase activity have been reported in both human (ROSEN et al. 1986) and rodent tumor cell lines (KANNER et al. 1988), although the significance of such an activation of *src* remains unclear. The central role of the *src* protein (and its close relatives) in growth regulation of normal and malignant cells dictates that we continue efforts to understand the structure and function of this important protein.

Acknowledgements. We thank our collaborators S. Parsons, M. Payne, A. Reynolds, S. Kanner, A. Bouton, W. Potts, W. Wasilenko, B. Cobb, R. Wang, D. Wages, and L. Kozma for permission to discuss unpublished data and for their help in preparation of this review. We also wish to thank our many colleagues for providing information prior to publication. The work carried out in the authors' laboratories was supported by Public Health Service grants CA40042, CA29243 and CA39076, and American Cancer Society grant NP462.

References

Alema S, Casaldore P, Agostini E, Tato F (1985) Differentiation of PC12 phaeochromocytoma cells induced by v-*src* oncogene. Nature 316: 557–559

Ali IV, Mautner V, Lanza R, Hynes RO (1977) Restoration of normal morphology adhesion and cytoskeleton in transformed cells by addition of a transformation-specific surface protein. Cell 11: 115–126

Allebach ES, Boettiger D, Pacifici M, Adams SL (1985) Control of types I and II collagen and fibronectin gene expression in chondrocytes delineated by viral transformation. Mol Cell Biol 5: 1002–1008

Ambros VR, Chen LB, Buchanan JM (1975) Surface ruffles as markers for studies of cell transformation by Rous sarcoma virus. Proc Natl Acad Sci USA 72: 3144–3148

Anderson DD, Beckmann RP, Harms EH, Nakamura K, Weber MJ (1981) Biological properties of "partial" transformation mutants of Rous sarcoma virus and characterization of their $pp60^{src}$ kinase. J Virol 37: 445–458

Anderson WB, Pastan I (1975) Altered adenylate cyclase activity: Its role in growth regulation and malignant transformation of fibroblasts. In: Drummond GI, Greengard P, Robison GA (eds) Advances in cyclic nucleotide research, vol 5. Raven, New York, pp 681–699

Antler A, Greenberg E, Edelman GM, Hanafusa H (1985) Increased phosphorylation of tyrosine in vinculin does not occur upon transformation by some avian sarcoma viruses. Mol Cell Biol 5: 263–267

Atkinson MM, Sheridan JD (1985) Reduced junctional permeability in cells transformed by different viral oncogenes. In: Bennett MVL, Spray DC (ed:) Gap Junctions. Cold Spring Laboratory, New York, pp 205–213

Azarnia R, Reddy S, Kmiecik TE, Shalloway D, Loewenstein WR (1988) The cellular *src* gene product regulates junctional cell-to-cell communication. Science 239: 398–401

Becker D, Kurth R, Critchley D, Friis R, Bauer H (1977) Distinguishable transformation-defective phenotypes among temperature-sensitive mutants of Rous sarcoma virus. J Virol 21: 1042–1055

Bedard P-A, Alcorta D, Simmons DL, Luk K-C, Erikson RL (1987) Constitutive expression of a gene encoding a polypeptide homologous to biologically active human platelet protein in Rous sarcoma virus-transformed fibroblasts. Proc Natl Acad Sci USA 84: 6715–6719

Beemon K, Ryden T, McNelly EA (1982) Transformation by avian sarcoma viruses leads to phosphorylation of multiple cellular proteins on tyrosine residues. J Virol 42: 742–747

Bishop JM (1987) The molecular genetics of cancer. Science 235: 305–311

Bolen JB, Thiele CJ, Israel MA, Yonemoto W, Lipsich LA, Brugge JS (1984) Enhancement of cellular *src* gene product associated tyrosyl kinase activity following polyoma virus infection and transformation. Cell 38: 767–777

Brenner S (1987) Phosphotransferase sequence homology. Nature 329: 21

Brugge JS (1986a) Interaction of the Rous sarcoma virus protein $pp60^{src}$ with the cellular proteins pp50 and pp90. Curr Top Microbiol Immunol 123: 1–22

Brugge JS (1986b) The p35/p36 substrates of protein-tyrosine kinases as inhibitors of phospholipase A_2. Cell 46: 149–150

Brugge JS, Darrow D (1984) Analysis of the catalytic domain of phosphotransferase activity of two avian sarcoma virus-transforming proteins. J Biol Chem 259: 4550–4557

Brugge JS, Erikson RL (1977) Identification of a transformation-specific antigen induced by an avian sarcoma virus. Nature 269: 346–348

Brugge JS, Erikson E, Erikson RL (1981) The specific interaction of the Rous sarcoma virus transforming protein, $pp60^{src}$, with two cellular proteins. Cell 25: 363–372

Brugge JS, Yonemoto W, Darrow D (1983) Interaction between the Rous sarcoma virus transforming protein and two cellular phosphoproteins: Analysis of the turnover and distribution of this complex. Mol Cell Biol 3: 9–19

Bryant D, Parsons JT (1982) Site-directed mutagenesis of the *src* gene of Rous sarcoma virus: Construction and characterization of a deletion mutant temperature sensitive for transformation. J Virol 44: 683–691

Bryant DL, Parsons JT (1984) Amino acid alterations within a highly conserved region of the Rous sarcoma virus *src* gene product $pp60^{src}$ inactivate tyrosine protein kinase activity. Mol Cell Biol 4: 862–866

Brzeski H, Ege T (1980) Changes in polypeptide pattern in ASV-transformed rat cells are correlated with the degree of morphological transformation. Cell 22: 513–522

Burr JG, Dreyfuss G, Penman S, Buchanan JM (1980) Association of the *src* gene product of Rous sarcoma virus with cytoskeletal structures of chicken embryo fibroblasts. Proc Natl Acad Sci USA 77: 3484–3488

Burr JG, Lee SR, Buchanan JM (1981) In situ phosphorylation of proteins associated with the cytoskeleton of chick embryo fibroblasts. In Cell Proliferation: Protein Phosphorylation, Vol. 8. ed: O. Rosen and E. Krebs. Cold Spring Harbor Press. pp 1217–1232

Buss JE, Sefton BM (1985) Myristic acid, a rare fatty acid, is the lipid attached to the transforming protein of Rous sarcoma virus and its cellular homolog. J Virol 53: 7–12

Buss JE, Kamps MP, Sefton BM (1984) Myristic acid is attached to the transforming protein of Rous sarcoma virus during or immediately after synthesis and is present in both soluble and membrane-bound forms of the protein, Mol Cell Biol 4: 2697–2704

Buss JE, Kamps MP, Gould K, Sefton BM (1986) The absence of myristic acid decreases membrane binding of $p60^{src}$ but does not affect tyrosine protein kinase activity. J Virol 58: 468–474

Calothy G, Poirier F, Dambrime G, Pessac B (1978) A transformation defective mutant of Rous sarcoma virus inducing chick embryo neuroretinal cell proliferation. Virology 89: 75–84

Calothy G, Laughier D, Cross FR, Jove R, Hanafusa T, Hanafusa H (1987) The membrane-binding domain and myristylation of $p60^{v\text{-}src}$ are not essential for stimulation of cell proliferation. J Virol 61: 1678–1681

Cartwright CA, Hutchinson MA, Eckhart W (1985) Structural and functional modification of $pp60^{v\text{-}src}$ associated with polyoma middle tumor antigen from infected or transformed cells. Mol Cell Biol 5: 2647–2652

Cartwright CA, Eckhart W, Simon S, Kaplan PL (1987a) Cell transformation by $pp60^{c\text{-}src}$ mutated in the carboxy-terminal regulatory domain. Cell 49: 83–91

Chackalaparampil I, Shalloway D (1988) Altered phosphorylation and activation of $pp60^{c\text{-}src}$ during fibroblast mitosis. Cell 52: 801–810

Chang C, Trosko JE, Kung H-J, Bombick D, Matsumura F (1985) Potential role of the *src* gene product in inhibition of gap-junctional communication in NIH/3T3 cells. Proc Natl Acad Sci USA 82: 5360–5364

Chang SH, Piwnica-Worms H, Harvey RW, Roberts TM, Smith AE (1988) The carboxy terminus of pp60$^{c\text{-}src}$ is a regulatory domain and is involved in complex formation with the middle-T antigen of polyomavirus. Mol Cell Biol 8: 1736–1747

Cheng YS, Chen LB (1981) Detection of phosphotyrosine-containing 34,000-dalton protein in the framework of cells transformed with Rous sarcoma virus. Proc Natl Acad Sci USA 78: 2388–2392

Collett MS, Erikson RL (1978) Protein kinase activity associated with the avian sarcoma virus *src* gene product. Proc Natl Acad Sci USA 75: 2021–2024

Collett MS, Erikson E, Erikson RL (1979) Structural analysis of the avian sarcoma virus transforming protein: Sites of phosphorylation. J Virol 29: 770–781

Comoglio PM, Di Renzo MF, Tarone G, Giancotti FG, Naldini L, Marchisio PC (1984) Detection of phosphotyrosine-containing proteins in the detergent-insoluble fraction of RSV-transformed fibroblasts by azobenzene phosphonate antibodies. EMBO J 3: 483–489

Cooper JA (1989) The *src* family of protein-tyrosine kinases. In: Kemp B, Alewood PF (ed) Peptides and protein phosphorylation. CRC Press, Boca Raton (in press)

Cooper JA, Hunter T (1981) Changes in protein phosphorylation in Rous sarcoma virus-transformed chicken embryo cells. Mol Cell Biol 1: 165–178

Cooper JA, MacAuley A (1988) Potential positive and negative autoregulation of p60$^{c\text{-}src}$ by intermolecular autophosphorylation. Proc Natl Acad Sci USA 85: 4232–4236

Cooper JA, Nakamura KD, Hunter T, Weber MJ (1983a) Phosphotyrosine-containing proteins and expression of transformation parameters in cells infected with partial transformation mutants of Rous sarcoma virus. J Virol 46: 15–28

Cooper JA, Reiss NA, Schwartz RJ, Hunter T (1983b) Three glycolytic enzymes are phosphorylated on tyrosine in cells transformed by Rous sarcoma virus. Nature 302: 218–222

Cooper JA, Esch FS, Taylor SS, Hunter T (1984a) Phosphorylation sites in enolase and lactate dehydrogenase utilized by tyrosine protein kinases in vivo and in vitro. J Biol Chem 259: 7835–7841

Cooper JA, Sefton BM, Hunter T (1984b) Diverse mitogenic agents induce the phosphorylation of two related 42,000-Dalton proteins on tyrosine in quiescent chick cells. Mol Cell Biol 4: 30–37

Cooper JA, Gould KL, Cartwright CA, Hunter T (1986) Tyr^{527} is phosphorylated in pp60$^{c\text{-}src}$: implications for regulation. Science 231: 1431–1434

Courtneidge SA (1985) Activation of the pp60$^{c\text{-}src}$ kinase by middle T antigen binding or by dephosphorylation. EMBO J 4: 1471–1477

Courtneidge SA, Bishop JM (1982) Transit of pp60$^{v\text{-}src}$ to the plasma membrane. Proc Natl Acad Sci USA 79: 7117–7121

Courtneidge SA, Heber A (1987) An 81 kd protein complexed with middle T antigen and pp60$^{c\text{-}src}$: a possible phosphatidylinositol kinase. Cell 50: 1031–1037

Courtneidge SA, Levinson AD, Bishop JM (1980) The protein encoded by the transforming gene of avian sarcoma virus (pp60^{src}) and a homologous protein in normal cells (pp60$^{proto\text{-}src}$) are associated with the plasma membrane. Proc Natl Acad Sci USA 77: 3783–3787

Courtneidge S, Ralston R, Alitalo K, Bishop JM (1982) Subcellular location of an abundant substrate (p36) for tyrosine-specific protein kinases. Mol Cell Biol 3: 340–350

Crompton MR, Moss SE, Crumpton MJ (1988) Diversity in the lipocortin/calpactin family. Cell 55: 1–3

Cross FR, Hanafusa H (1983) Local mutagenesis of Rous sarcoma virus: the major sites of tyrosine and serine phosphorylation of p60^{src} are dispensable for transformation. Cell 34: 597–607

Cross FR, Garber EA, Pellman D, Hanafusa H (1984) A short sequence in the p60^{src} N terminus is required for p60^{src} myristylation and membrane association and for cell transformation. Mol Cell Biol 4: 1834–1842

Cross FR, Garber EA, Hanafusa H (1985) N-terminal deletions in Rous sarcoma virus p60^{src}: effects on tyrosine kinase and biological activities and on recombination in tissue culture with the cellular *src* gene. Mol Cell Biol 5: 2789–2795

DeClue JE, Martin GS (1987) Phosphorylation of talin at tyrosine in Rous sarcoma virus-transformed cells. Mol Cell Biol 7: 371–378

De Clue JE, Martin GS (1989) Linker insertion/deletion mutagenesis of the v-*src* gene: isolation of host- and temperature-dependent mutants. J Virol 63: 542–554

DeClue JE, Sadowski I, Martin GS, Pawson T (1987) A conserved domain regulates interactions of the v-*fps* protein-tyrosine kinase with the host cell. Proc Natl Acad Sci USA 84: 9064–9068

de Crombrugghe B, Schmidt A, Liau G, Setoyama C, Mudryj M, Yamada Y, McKeon C (1985) Structural and functional analysis of the genes for α(I) and α(III) collagens. Ann NY Acad Sci 460: 154–162

Deichaite I, Casson LP, Ling H-P, Resh MD (1988) In vitro synthesis of pp60$^{v\text{-}src}$: myristylation in a cell-free system. Mol Cell Biol 8: 4295–4301

Ferrell JE, Martin GS (1988) Platelet tyrosine-specific protein phosphorylation is regulated by thrombin. Mol Cell Biol 8: 3603–3610

Fincham VJ, Wyke JA (1986) Localization of temperature-sensitive transformation mutations and back mutations in Rous sarcoma virus *src* gene. J Virol 58: 694–699

Fincham VJ, Chiswell DJ, Wyke JA (1982) Mapping of nonconditional and conditional mutants in the *src* gene of Prague strain Rous sarcoma virus. Virology 116: 72–83

Flier JS, Mueckler MM, Usher P, Lodish HF (1987) Elevated levels of glucose transport and transporter messenger RNA are induced by *ras* or *src* oncogenes. Science 235: 1492–1498

Fujita DJ, Bechberger J, Nedic I (1981a) Four Rous sarcoma virus mutants which affect transformed cell morphology exhibit altered *src* gene products. Virology 114: 256–260

Fujita DJ, Boschek CB, Ziemiecki A, Friis RR (1981b) An avian sarcoma virus mutant which produces an aberrant transformation affecting cell morphology. Virology 111: 223–238

Fukami Y, Nakamura T, Nakayama A, Kanehisa T (1986) phosphorylation of tyrosine residues of calmodulin in Rous sarcoma virus-transformed cells. Proc Natl Acad Sci USA 83: 4190–4193

Gacon G, Gisselbrecht S, Piau J-P, Fiszman M-Y, Fischer S (1982) Phosphorylation of the subcellular matrix in cells transformed by Rous' sarcoma virus. Eur. J. Biochem 125: 453–456

Garber EA, Hanafusa H (1987) NH_2-terminal sequences of two *src* protein that cause aberrant transformation. Proc Natl Acad Sci USA 84: 80–84

Garber EA, Krueger JG, Hanafusa H, Goldberg AR (1983) Temperature-sensitive membrane association of pp60^{src} in tsNY68-infected cells correlates with increased tyrosine phosphorylation of membrane-associated proteins. Virology 126: 73–86

Garber EA, Cross FR, Hanafusa H (1985) Processing of p60$^{v\text{-}src}$ to its myristylated membrane-bound form. Mol Cell Biol 5: 2781–2788

Garber EA, Mayer BJ, Jove R, Hanafusa H (1987) Analysis of p60$^{v\text{-}src}$ mutants carrying lesions involved in temperature sensitivity. J Virol 61: 354–360

Golden A, Nemeth SP, Brugge JS (1986) Blood platelets express high levels of the pp60$^{c\text{-}src}$-specific tyrosine kinase activity. Proc Natl Acad Sci USA 83: 852–856

Gould KL, Hunter T (1988) Platelet-derived growth factor induces multisite phosphorylation of pp60$^{c\text{-}src}$ and increases its protein-tyrosine kinase activity. Mol Cell Biol 8: 3345–3356

Gould KL, Woodgett JR, Cooper JA, Buss JE, Shalloway D, Hunter T (1985) Protein kinase C phosphorylates pp60^{src} at a novel site. Cell 42: 849–857

Gould KL, Cooper JA, Bretscher A, Hunter T (1986a) The protein-tyrosine kinase substrate, p81, is homologous to a chicken microvillar core protein. J Cell Biol 102: 660–669

Gould KL, Woodgett JR, Isacke CM, Hunter T (1986b) The protein-tyrosine kinase substrate p36 is also a substrate for protein kinase C in vitro and in vivo. Mol Cell Biol 6: 2738–2744

Greenberg ME, Edelman GM (1983) The 34 kd pp60^{src} substrate is located at the inner face of the plasma membrane. Cell 33: 767–779

Groudine M, Weintraub H (1975) Rous sarcoma virus activate embryonic globin genes in chicken fibroblasts. Proc Natl Acad Sci USA 72: 4464–4468

Hamaguchi M, Hanafusa H (1987) Association of p60^{src} with triton-resistant cellular structure correlates with morphological transformation. Proc Nat Acad Sci USA 84: 2312–2316

Hamaguchi M, Grandori C, Hanafusa H (1988) Phosphorylation of cellular proteins in Rous sarcoma virus-infected cells: Analysis by use of anti-phosphotyrosine antibodies. Mol Cell Biol 8: 3035–3042

Hanks SK, Quinn AM, Hunter T (1988) The protein kinase family: conserved features and deduced phylogeny of the catalytic domains. Science 241: 42–52

Hendricks M, Weintraub H (1984) Multiple tropomyosin polypeptides in chicken embryo fibroblasts: differential repression of transcription by Rous sarcoma virus transformation. Mol Cell Biol 4: 1823–1833

Hirota Y, Kato J-Y, Takeya T (1988) Substitution of Ser-17 of pp60$^{c\text{-}src}$: biological and biochemical characterization in chicken embryo fibroblasts. Mol Cell Biol 8: 1826–1830

Hirst R, Horwitz A, Buck C, Rohrschneider L (1986) Phosphorylation of the fibronectin receptor complex in cells transformed by oncogenes that encode tyrosine kinases. Proc Natl Acad Sci USA 83: 6470–6474

Hunter T, Cooper JA (1985) Protein-tyrosine kinases. Annu Rev Biochem 54: 897–930

Hunter T, Sefton BM (1980) Transforming gene product of Rous sarcoma virus phosphorylates tyrosine. Proc Natl Acad Sci USA 77: 1311–1315

Iba H, Takeya T, Cross FR, Hanafusa T, Hanafusa H (1984) Rous sarcoma virus variants which carry the cellular *src* gene instead of the viral *src* gene cannot transform chicken embryo fibroblasts. Proc Natl Acad Sci USA 81: 4424–4428

Iwashita S, Kitamura N, Yoshida M (1983) Molecular events leading to fusiform morphological transformation by partial *src* deletion mutant of Rous sarcoma virus. Virology 125: 419–431

Jove R, Hanafusa H (1987) Cell transformation by the viral *src* oncogene. Annu Rev Cell Biol 3: 31–56

Jove R, Garber EA, Iba H, Hanafusa H (1986a) Biochemical properties of p60$^{v\text{-}src}$ mutants that induce different cell transformation parameters. J Virol 60: 849–857

Jove R, Mayer BJ, Iba H, Laugier D, Poirier F, Calothy G, Hanafusa T, Hanafusa H (1986b) Genetic analysis of p60$^{v\text{-}src}$ domains involved in the induction of different cell transformation parameters. J Virol 60: 840–848

Jove R, Kornbluth S, Hanafusa H (1987) Enzymatically inactive p60$^{c\text{-}src}$ mutant with altered ATP-binding site is fully phosphorylated in its carboxy-terminal regulatory region. Cell 50: 937–943

Kahn P, Nakamura K, Shin S, Smith RE, Weber MJ (1982) Tumorigenicity of partial transformation mutants of Rous sarcoma virus. J Virol 42: 602–611

Kaji A, Amini S (1983) Association of pp36, a phosphorylated form of the presumed target protein for the *src* protein of Rous sarcoma virus, with the membrane of chicken cells transformed by Rous sarcoma virus. Proc Natl Acad Sci USA 80: 960–964

Kamps MP, Sefton BM (1986) Neither arginine nor histidine can carry out the function of lysine-295 in the ATP-binding site of p60^{src}. Mol Cell Biol 6: 751–757

Kamps MP, Sefton BM (1988) Identification of multiple novel polypeptide substrates of the v-*src*, v-*yes*, v-*fps*, *v-ros*, and v-*erb*-B oncogenic tyrosine protein kinases utilizing antisera against phosphotyrosine. Oncogene 2: 305–315

Kamps MP, Taylor SS, Sefton BM (1984) Direct evidence that oncogenic tyrosine kinases and cyclic AMP-dependent protein kinase have homologous ATP-binding sites. Nature 310: 589–592

Kamps MP, Buss JE, Sefton BM (1985) Mutation of NH_2-terminal glycine of p60^{src} prevents both myristoylation and morphological transformation. Proc Natl Acad Sci USA 82: 4625–4628

Kamps MP, Buss JE, Sefton BM (1986) Rous sarcoma virus transforming protein lacking myristic acid phosphorylates known polypeptide substrates without inducing transformation. Cell 45: 105–112

Kanner SB, Parsons SJ, Parsons TJ, Gilmer TM (1988) Activation of pp60$^{c\text{-}src}$ tyrosine kinase specific activity in tumor-derived Syrian hamster embryo cells. Oncogene 2: 327–335

Kaplan DR, Whitman, Schaffhausen, Pallas DC, White M, Cantley L, Roberts TM (1987) Common elements in growth factor stimulation and oncogenic transformation: 85 kd phosphoprotein and phosphatidylinositol kinase activity. Cell 50: 1021–1029

Kaplan JM, Mardon G, Bishop JM, Varmus HE (1988) The first seven amino acids encoded by the v-*src* oncogene act as a myristylation signal: lysine 7 is a critical determinant. Mol Cell Biol 8: 2435–2441

Kato J-Y, Takeya T, Grandori C, Iba H, Levy JB, Hanafusa H (1986) Amino acid substitutions sufficient to convert the nontransforming p60$^{c\text{-}src}$ protein to a transforming protein. Mol Cell Biol 6: 4155–4160

Kazlauskas A, Cooper JA (1988) Protein kinase C mediates platelet-derived growth factor-induced tyrosine phosphorylation of p42. J Cell Biol 106: 1395–1402

Kawai S, Hanafusa H (1971) The effects of reciprocal changes in temperature on the transformed state of cells infected with a Rous sarcoma virus mutant. Virology 46: 470–479

Kellie S, Patel B, Mitchell A, Critchley DR, Wigglesworth NM, Wyke JA (1986a) Comparison of the relative importance of tyrosine-specific vinculin phosphorylation and the loss of surface-associated fibronectin in the morphology of cells transformed by Rous sarcoma virus. J Cell Sci 82: 129–142

Kellie S, Patel B, Wigglesworth NM, Critchley DR, Wyke JA (1986b) The use of Rous sarcoma virus transformation mutants with differing tyrosine kinase activities to study the relationships between vinculin phosphorylation, $pp60^{v\text{-}src}$ location and adhesion plaque integrity. Exp Cell Res 165: 216–228

Kitamura N, Yoshida M (1983) Small deletion in *src* of Rous sarcoma virus modifying transformation phenotypes: identification of 207-nucleotide deletion and its smaller product with protein kinase activity. J Virol 46: 985–992

Kmiecik TE, Shalloway D (1987) Activation and suppression of $pp60^{c\text{-}src}$ transforming ability by mutation of its primary site of tyrosine phosphorylation. Cell 49: 65–73

Kmiecik TE, Johnson PJ, Shalloway D (1988) Regulation by the autophosphorylation site in overexpressed $pp60^{c\text{-}src}$. Mol Cell Biol 8: 4541–4546

Krueger JG, Garber EA, Goldberg AR, Hanafusa H (1982a) Changes in amino-terminal sequences of $pp60^{src}$ lead to decreased membrane association and decreased in vivo tumorigenicity. Cell 28: 889–896

Krueger JG, Garber EA, Goldberg AR, Hanafusa H (1982b) Evidence that the *src* gene product of Rous sarcoma virus is membrane associated. Virology 101: 25–40

Krueger JG, Garber EA, Goldberg AR (1983) Subcellular localization of $pp60^{src}$ in RSV-transformed cells. Curr Top Microbiol Immunol 107: 52–124

Krueger JG, Garber EA, Chin SS-M, Hanafusa H, Goldberg AR (1984) Size-variant $pp60^{src}$ proteins of recovered avian sarcoma virus interact with adhesion plaques as peripheral membrane proteins: effects on cell transformation. Mol Cell Biol 4: 454–467

Lau AF (1986) Phosphotyrosine-containing 120,000-dalton protein coimmunoprecipitated with $pp60^{v\text{-}src}$ from Rous sarcoma virus-transformed mammalian cells. Virology 151: 86–99

Laudano AP, Buchanan JM (1986) Phosphorylation of tyrosine in the carboxyl-terminal tryptic peptide of $pp60^{c\text{-}src}$. Proc Natl Acad Sci USA 83: 892–896

Levinson AD, Oppermann H, Levintow L, Varmus HE, Bishop JM (1978) Evidence that the transforming gene of avian sarcoma virus encodes a protein kinase associated with a phosphoprotein. Cell 15: 561–572

Levy JB, Iba H, Hanafusa H (1986) Activation of the transforming potential of $p60^{c\text{-}src}$ by a single amino acid change. Proc Natl Acad Sci USA 83: 4228–4232

Levy JB, Dorai T, Wang L-H, Brugge JS (1987) The structurally distinct form of $pp60^{c\text{-}src}$ detected in neuronal cells is encoded by a unique c-*src* mRNA. Mol Cell Biol 7: 4142–4145

Linder ME, Burr JG (1988a) Immunological characterization of proteins detected by phosphotyrosine antibodies in cells transformed by Rous sarcoma virus. J Virol 62: 2665–2673

Linder ME, Burr JG (1988b) Nonmyristoylated $p60^{v\text{-}src}$ fails to phosphorylate proteins of 115–120 kDa in chicken embryo fibroblasts. Proc Natl Acad Sci USA 85: 2608–2612

Luttrell DK, Luttrell LM, Parsons SJ (1988) Augmented mitogenic responsiveness to epidermal growth factor in murine fibroblasts that overexpress $pp60^{c\text{-}src}$. Mol Cell Biol 8: 497–501

Mardon G, Varmus HE (1983) Frameshift and intragenic suppressor mutations in a Rous sarcoma provirus suggest *src* encodes two proteins. Cell 32: 871–879

Martin GS (1970) Rous sarcoma virus: a function required for the maintenance of the transformed state. Nature 227: 1021–1023

Martin GS, Duesberg PH (1972) The a subunit in the RNA of transforming avian tumor viruses. I. Occurrence in different virus strains. II. Spontaneous loss resulting in nontransforming variants. Virology 47: 494–497

Martinez R, Nakamura KD, Weber MJ (1982) Identification of phosphotyrosine-containing proteins in untransformed and Rous sarcoma virus-transformed chicken embryo fibroblasts. Mol Cell Biol 2: 653–665

Martinez R, Mathey-Prevot B, Bernards A, Baltimore D (1987) Neuronal $pp60^{c\text{-}src}$ contains a six amino acid insertion relative to its non-neuronal counterpart. Science 237: 411–415

Mayer B, Jove R, Krane JF, Poirier F, Calothy G, Hanafusa H (1986) Genetic lesions involved in temperature sensitivity of the *src* gene products of four Rous sarcoma virus mutants. J Virol 60: 858–867

Mayer BJ, Hamaguchi M, Hanafusa H (1988) A novel viral oncogene with structural similarity to phospholipase C. Nature 332: 272–275

MacAuley A, Cooper JA (1988) The carboxy-terminal sequence of $p56^{lck}$ can regulate $p60^{c\text{-}src}$. Mol Cell Biol 8: 3560–3564

Monteagudo CA, Williams DL, Crabb GA, Tondravi M, Weber MJ (1984) Phosphotyrosine-containing membrane proteins in Rous sarcoma virus-transformed cells. Cold Spring Harbor Meetings on Cell Proliferation: The Cancer Cell 2: 69–75

Nakamura KD, Weber MJ (1982) Phosphorylation of a 36,000 M_r cellular protein in cells infected with partial transformation mutants of Rous sarcoma virus. Mol Cell Biol 2: 147–153

Nakamura KD, Martinez R, Weber MJ (1983) Tyrosine phosphorylation of specific proteins after mitogen stimulation of chicken embryo fibroblasts. Mol Cell Biol 3: 380–390

Nemeth SP, Fox LG, DeMarco M, Brugge JS (1989) Deletions within the amino-terminal half of the c-*src* gene product which alter the functional activity of the protein. Mol Cell Biol 9: 1109–1119

Nishizawa M, Mayer BJ, Takeya T, Yamamoto T, Toyoshima K, Hanafusa H, Kawai S (1985) Two independent mutations are required for temperature-sensitive cell transformation by a Rous sarcoma virus temperature-sensitive mutant. J Virol 56: 743–749

Oppermann H, Levinson AD, Levintow L, Varmus HE, Bishop JM, Kawai S (1981) Two cellular proteins that immunoprecipitate with the transforming protein of Rous sarcoma virus. Virology 113: 736–751

Parker RC, Varmus HE, Bishop JM (1984) Expression of v-*src* and chicken c-*src* in rat cells demonstrates qualitative differences between $pp60^{v\text{-}src}$ and $pp60^{c\text{-}src}$. Cell 37: 131–139

Parsons SJ, Creutz CE (1986) $p60^{c\text{-}src}$ activity detected in the chromaffin granule membrane. Biochem Biophys Res Commun 134: 736–742

Pasquale EB, Maher PA, Singer SJ (1986) Talin is phosphorylated on tyrosine in chicken embryo fibroblasts transformed by Rous sarcoma virus. Proc Natl Acad Sci USA 83: 5507–5511

Patschinsky T, Hunter T, Esch FS, Cooper JA, Sefton BM (1982) Analysis of the sequence of amino acids surrounding sites of tyrosine phosphorylation. Proc Natl Acad Sci USA 79: 973–977

Patschinsky T, Hunter T, Sefton BM (1986) Phosphorylation of the transforming protein of Rous sarcoma virus: Direct demonstration of phosphorylation of serine 17 and identification of an additional site of tyrosine phosphorylation in $p60^{v\text{-}src}$ of Prague Rous sarcoma virus. J Virol 59: 73–81

Pellman D, Garber EA, Cross FR, Hanafusa H (1985a) An N-terminal peptide from $p60^{src}$ can direct myristylation and plasma membrane localization when fused to heterologous proteins. Nature 314: 374–377

Pellman D, Garber EA, Cross FR, Hanafusa H (1985b) Fine structural mapping of a critical NH_2-terminal region of $p60^{src}$. Proc Natl Acad Sci USA 82: 1623–1627

Piwnica-Worms H, Saunders KB, Roberts TM, Smith AE, Cheng SH (1987) Tyrosine phosphorylation regulates the biochemical and biological properties of $pp60^{c\text{-}src}$. Cell 49: 75–82

Potts WM, Reynolds AB, Lansing TJ, Parsons JT (1988) Activation of $pp60^{c\text{-}src}$ transforming potential by mutations altering the structure of an amino terminal domain containing residues 90–95. Oncogene Res 3: 343–355

Purchio AF (1982) Evidence that $pp60^{src}$, the product of the Rous sarcoma virus *src* gene, undergoes autophosphorylation. J Virol 41: 1–7

Purchio AF, Shoyab M, Gentry LE (1985) Site-specific increased phosphorylation of $pp60^{v\text{-}src}$ after treatment of RSV-transformed cells with a tumor promoter. Science 229: 1393–1395

Radke K, Martin GS (1979) Transformation by Rous sarcoma virus: Effects of *src* gene expression on the synthesis and phosphorylation of cellular polypeptides. Proc Natl Acad Sci 76: 5212–5216

Radke K, Gilmore T, Martin GS (1980) Transformation by Rous sarcoma virus: a cellular substrate for transformation-specific protein phosphorylation contains phosphotyrosine. Cell 21: 821–828

Radke K, Carter VC, Moss P, Dehazya P, Schliwa M, Martin GS (1983) Membrane association of a 36,000-dalton substrate for tyrosine phosphorylation in chicken embryo fibroblasts transformed by avian sarcoma viruses. J Cell Biol 97: 1601–1611

Ralston R, Bishop JM (1985) The product of the protooncogene c-*src* is modified during the cellular response to platelet-derived growth factor. Proc Natl Acad Sci USA 82: 7845–7849
Raymond VW, Parsons JT (1987) Identification of an amino terminal domain required for the transforming activity of the Rous sarcoma virus *src* protein. Virology 160: 400–410
Rees-Jones RW, Goff SP (1988) Insertional mutagenesis of the Abelson murine leukemia virus genome: identification of mutants with altered kinase activity and defective transformation ability. J Virol 62: 978–986
Resh MD (1988) Reconstitution of the Rous sacroma virus transforming protein $pp60^{v\text{-}src}$ into phospholipid vesicles. Mol Cell Biol 8: 1896–1905
Resh MD, Erikson RL (1985a) Highly specific antibody to Rous sarcoma virus *src* gene product recognizes a novel population of $pp60^{v\text{-}src}$ and $pp60^{c\text{-}src}$ molecules. J Cell Biol 100: 409–417
Resh MD, Erikson RL (1985b) Characterization of $pp60^{src}$ phosphorylation in vitro in Rous sarcoma virus-transformed cell membranes. Mol Cell Biol 5: 916–922
Reynolds AB, Vila J, Lansing TJ, Potts WM, Weber MJ, Parsons JT (1987) Activation of the oncogenic potential of the avian cellular *src* protein by specific structural alteration of the carboxy terminus. EMBO J 6: 2359–2364
Reynolds AB, Roesel DJ, Kanner SB, Parsons JT (1989) Transformation-specific tyrosine phosphorylation of a novel cellular protein in chicken cells expressing oncogenic variants of the avian cellular *src* gene. Mol Cell Biol 19: 629–638
Reynolds AB, Kanner SB, Wang H-C, Parsons JT (1989) The stable association of activated $pp60^{src}$ with two tyrosine phosphorylated cellular proteins. Mol Cell Biol 9: in press
Rifkin DB, Beal LP, Reich E (1975) Macromolecular determinants of plasminogen activator synthesis. In: Reich E, Rifkin DB, Shaw E (eds) Proteases and biological control. Cold Spring Harbor Laboratory, New York, pp 841–847
Rohrschneider L (1980) Adhesion plaques of Rous sarcoma virus-transformed cells contain the *src* gene product. Proc Natl Acad Sci USA 77: 3514–3518
Rohrschneider L, Reynolds S (1985) Regulation of cellular morphology by the Rous sarcoma virus *src* gene: analysis of fusiform mutants. Mol Cell Biol 5: 3097–3107
Rohrschneider L, Rosok MJ (1983) Transformation parameters and $pp60^{src}$ localization in cells infected with partial transformation mutants of Rous sarcoma virus. Mol Cell Biol 3: 731–746
Rosen N, Bolen JB, Schwartz AM, Cohen P, DeSeau V, Israel MA (1986) Analysis of $pp60^{c\text{-}src}$ protein kinase activity in human tumor cell lines and tissues. J Biol Chem 261: 13754–13759
Roth CW, Richert ND, Pastan I, Gottesman MM (1983) Cyclic AMP treatment of Rous sarcoma virus-transformed Chinese hamster ovary cells increases phosphorylation of $pp60^{src}$ and increases $pp60^{c\text{-}src}$ kinase activity. J Biol Chem 258: 10768–10773
Rous P (1910) An experimental comparison of transplanted tumor and a transplanted normal tissue capable of growth. J Exp Med 12: 344–366
Sadowski I, Stone JC, Pawson T (1986) A noncatalytic domain conserved among cytoplasmic protein-tyrosine kinases modifies the kinase function and transforming activity of Fujinami sacroma virus $P130^{gag\text{-}fps}$. Mol Cell Biol 6: 4396–4408
Schultz AM, Henderson LE, Oroszlan S, Garber EA, Hanafusa H (1985) Amino terminal myristylation of the protein kinase $p60^{src}$, a retroviral transforming protein. Science 227: 427–429
Schwartz DE, Tizard R, Gilbert W (1983) Nucleotide sequence of Rous sarcoma virus. Cell 32: 853–869
Sefton BM, Hunter T, Beemon K, Eckhart W (1980) Evidence that the phosphorylation of tyrosine is essential for cellular transformation by Rous sarcoma virus. Cell 20: 807–816
Sefton BM, Hunter T, Ball EH, Singer SJ (1981) Vinculin: a cytoskeletal target of the transforming protein of Rous sarcoma virus. Cell 24: 165–174
Shalloway D, Coussens PM, Yaciuk P (1984) c-*src* and *src* homolog overexpression in mouse cells. In: Vande Woude GF (ed) Cancer Cells 2: Oncogenes and viral genes. Cold Spring Harbor Laboratory, New York, pp 9–17
Smart JE, Oppermann H, Czernilofsky AP, Purchio AF, Erikson RL, Bishop JM (1981) Characterization of sites for tyrosine phosphorylation in the transforming protein of Rous sarcoma virus ($pp60^{v\text{-}src}$) and its normal cellular homologue ($pp60^{c\text{-}src}$). Proc Natl Acad Sci USA 78: 6013–6017
Snyder MA, Bishop JM (1984) A mutation at the major phosphotyrosine in $pp60^{v\text{-}src}$ alters oncogenic potential. J Virol 136: 375–386

Snyder MA, Bishop JM, Colby WW, Levinson AD (1983) Phosphorylation of tyrosine-416 is not required for the transforming properties and kinase activity of pp60$^{v\text{-}src}$. Cell 32: 891–901

Snyder MA, Bishop JM, McGrath JP, Levinson AD (1985) A mutation at the ATP-binding site of pp60$^{v\text{-}src}$ abolishes kinase activity, transformation, and tumorigenicity. Mol Cell Biol 5: 1772–1779

Stahl ML, Ferenz CR, Kelleher KL, Kriz RW, Knopf JL (1988) Sequence similarity of phospholipase C with the non-catalytic region of *src*. Nature 332: 269–272

Stehelin D, Varmus HE, Bishop JM, Vogt PK (1976) DNA related to the transforming gene(s) of avian sarcoma viruses is present in normal avian DNA. Nature 260: 170–173

Sternberg MJE, Taylor WR (1984) Modelling the ATP-binding site of oncogene products, the epidermal growth factor receptor and related proteins. FEBS 175: 387–392

Stoker AW, Enrietto PJ, Wyke JA (1984) Functional domain of the pp60$^{v\text{-}src}$ protein as revealed by analysis of temperature-sensitive Rous sarcoma virus mutants. Mol Cell Biol 4: 1508–1514

Stoker AW, Kellie S, Wyke JA (1986) Intracellular localization and processing of pp60$^{v\text{-}src}$ proteins expressed by two distinct temperature-sensitive mutants of Rous sarcoma virus. J Virol 58: 876–883

Stone JC, Atkinson T, Smith M, Pawson T (1984) Identification of functional regions in the transforming protein of Fujinami sarcoma virus by in-phase insertion mutagenesis. Cell 37: 549–558

Sugano S, Stoeckle MY, Hanafusa H (1987) Transformation by Rous sarcoma virus induces a novel gene with homology to a mitogenic platelet protein. Cell 49: 321–328

Suh P-G, Ryo SH, Moon KH, Suh HW, Rhee SG (1988) Cloning and sequence of multiple forms of phospholipase C. Cell 54: 161–169

Takeya T, Hanafusa H (1983) Structure and sequence of the cellular gene homologous to the RSV *src* gene and the mechanism for generating the transforming virus. Cell 32: 881–890

Tapley P, Horwitz A, Buck C, Burrige K, Duggan K, Rohrschneider L (1989) Integrins isolated from Rous sarcoma virus-transformed chicken embryo fibroblasts. Oncogene, 4: 325–334

Temin HM (1960) The control of cellular morphology in embryonic cells infected with Rous sarcoma virus in vitro. Virology 10: 182–197

Temin HM (1962) Separation of morphological conversion and virus production in Rous sarcoma virus infection. Cold Spring Harbor Sym Quant Biol 27: 407–414

Toyoshima K, Vogt PK (1969) Temperature sensitive mutants of an avian sarcoma virus. Virology 39: 930–931

Tyagi JS, Hirano H, Pastan I (1985) Modulation of fibronectin gene activity in chick embryo fibroblasts transformed by a temperature-sensitive strain (ts68) of Rous sarcoma virus. Nucleic Acid Res 13: 8275–8284

Verderame MF, Kaplan JM, Varmus HE (1989) A mutation in v-*src* removing a single conserved residue in the SH2 domain of pp60 restricts transformation in a host-dependent manner. J Virol 63: 338–348

Vila J, Weber MJ (1988) Mitogen-stimulated phosphorylation of a 42-kD cellular protein: Evidence for a protein kinase-C requirement. J Cell Physiol 135: 285–292

Vogel US, Dixon RAF, Schaber MD, Diehl RE, Marshall MS, Scolnick IS, Gibbs JB (1988) Cloning of bovine GAP: a protein that interacts with oncogenic *ras* p21. Nature 335: 90–93

Vogt PK (1971) Spontaneous segregation of nontransforming viruses from cloned sarcoma viruses. Virology 46: 939–946

Wang E, Goldberg AR (1976) Changes in microfilament organization and surface topography upon transformation of chick embryo fibroblasts with Rous sarcoma virus. Proc Natl Acad Sci USA 73: 4065–4069

Wang H-CR, Parsons JT (1989) Deletions and insertions within an amino terminal domain of pp60$^{v\text{-}src}$ inactivate transformation and modulate membrane stability. J Virol 63: 291–302

Wasilenko WJ, Shawver LK, Weber MJ (1987) Down-modulation of EGF receptors in cells transformed by the *src* oncogene. J Cell Physiol 131: 450–457

Weber MJ (1974) Reversal of the transformed phenotype by dibutyryl cyclic AMP and a protease inhibitor. In: Fox CF, Robinson WW (eds) Proceedings of the conference on mechanisms of virus disease. Benjamin, Menlo Park, New York, pp 327–345

Weber MJ (1984) Malignant transformation by Rous sarcoma virus: from phosphorylation to phenotype. Adv Viral Oncology 4: 249–268

Weber MJ, Friis RR (1979) Dissociation of transformation parameters using temperature-conditional mutants of Rous sarcoma virus. Cell 16: 25–32

Welham MJ, Wyke JA (1988) A single point mutation has pleiotropic effects on $pp60^{v\text{-}src}$ function. J Virol 62: 1898–1906

White MK, Weber MJ (1988) Transformation by the *src* oncogene alters glucose transport into rat and chicken cells by different mechanisms. Mol Cell Biol 8: 138–144

Wierenga RK, Hol WGJ (1983) Predicted nucleotide-binding properties of p21 protein and its cancer-associated variant. Nature 302: 842–844

Wilcox C, Hu J-S, Olson EN (1987) Acylation of proteins with myristic acid occurs cotranslationally. Science 238: 1275–1278

Wilkerson VW, Bryant DL, Parsons JT (1985) Rous sarcoma virus variants that encode *src* proteins with an altered carboxy terminus are defective for cellular transformation. J Virol 55: 314–321

Willingham MC, Jay C, Pastan I (1979) Localization of the ASV *src* gene product to the plasma membrane of transformed cells by electron microscopic immunocytochemistry. Cell 18: 125–134.

Wilson LK, Luttrell DK, Parsons JT, Parsons SJ (1989) $pp60^{c\text{-}src}$ tyrosine kinase, myristylation, and modulatory domains are required for the enhanced mitogenic responsiveness to epidermal growth factor seen in cells overexpressing c-*src*. Mol Cell Biol 9: 1536–1544

Wyke JA, Stoker AW (1987) Genetic analysis of the form and function of the viral *src* oncogene product. Biochim Biophys Acta 907: 47–69

Yaciuk P, Shalloway D (1986) Features of the $pp60^{v\text{-}src}$ carboxyl terminus that are required for transformation. Mol Cell Biol 6: 2807–2819

Yaciuk P, Cannella MT, Shalloway D (1988) Comparison of the effects of carboxyl terminal truncation and point mutations on $pp60^{c\text{-}src}$ activities. Oncogene Res 3: 207–212

Yonemoto W, Jarvis-Morar M, Brugge JS, Bolen JB, Israel MA (1985) Tyrosine phosphorylation within the amino-terminal domain of $pp60^{c\text{-}src}$ molecules associated with polyoma virus middle-sized tumor antigen. Proc Natl Acad Sci USA 82: 4568–4572

Yonemoto W, Filson AJ, Queral-Lustig AE, Wang JYJ, Brugge JS (1987) Detection of phosphotyrosine-containing proteins in polyomavirus middle tumor antigen-transformed cells after treatment with a phosphotyrosine phosphatase inhibitor. Mol Cell Biol 7: 905–913

Structures and Activities of Activated *abl* Oncogenes

REX RISSER and GINA D. HOLLAND

1 Introduction

Activated versions of the c-*abl* proto-oncogene have been recovered in transforming retroviruses of mice and cats, and have been recovered from various forms of the 9:22 (Philadelphia) chromosomal translocation found in some human leukemias. The purpose of this review is to summarize what is known about the various structures of activated *abl* oncogenes, what is known about the proteins encoded by the *abl* proto-oncogene and its activated derivatives, and how domains and biochemical functions ascribed to *abl* proteins relate to those of other members of the tyrosine kinase family of enzymes. Finally, the far less defined activities of *abl* proteins in mediating cell transformation and tumor induction will be reviewed.

Like the classical reactions of organic chemistry, retroviruses with unique biological properties frequently have been named for their discoverers, in the present case Abelson murine leukemia virus, A-MuLV (ABELSON and RABSTEIN 1970a, b).

McArdle Laboratory for Cancer Research, University of Wisconsin, Madison, WI 53706, USA

Current Topics in Microbiology and Immunology, Vol. 147

A-MuLV was originally recovered as a pathogenic variant of Moloney MuLV. It differed from M-MuLV in its rapid ability to induce non-thymic lymphosarcoma, its apparent inability to induce thymic lymphoma, and its ability to induce neoplastic disease following infection of adult mice (ABELSON and RABSTEIN 1970b; SIEGLER et al. 1972; RISSER et al. (1978). It is now clear that all of these properties can be attributed to a unique gene that A-MuLV encodes, i.e., the v-*abl* oncogene (WITTE et al. 1978, 1980; REYNOLDS et al. 1978; GOFF et al. 1982; ENGLEMAN and ROSENBERG 1987; GREEN et al. 1987; KIPREOS et al. 1987). Although the recombination event which generated A-MuLV was recovered only once, a similar event involving the retroviral transduction of the c-*abl* proto-oncogene has been documented in cats (BESMER et al. 1983). Events with similar consequences for the function of the c-*abl* protein occur frequently in humans, probably in each person who develops chronic myelogenous leukemia (CML) (DE KLEIN et al. 1982; HEISTERKAMP et al. 1983; CANAANI et al. 1984; KONOPKA et al. 1984; DAVIS et al. 1985; SHTIVELMAN et al. 1985) and in many who develop acute lymphoblastic leukemia (ALL) (HERMANS et al. 1987).

2 The Structures of Activated *abl* Oncogenes and Their Corresponding Messenger RNAs

The A-MuLV genome was generated by recombination between M-MuLV and a processed form of the c-*abl* proto-oncogene (Fig. 1). As its name implies, the c-*abl* proto-oncogene is the cellular gene from which activated *abl* oncogenes are derived following transduction by retroviruses (GOFF et al. 1980; WANG et al. 1984) or chromosomal translocation (DE KLEIN et al. 1982). Homologs of the c-*abl* proto-oncogene have been characterized in fruit flies (HOFFMANN et al. 1983; HENKEMEYER et al. 1988) and nematodes (GODDARD et al. 1986). Genetic experiments in flies indicate that the fly c-*abl* gene plays an essential role in development (HENKEMEYER et al. 1987; see also HOFFMANN, this volume). The c-*abl* proto-oncogene spans approximately 250 kilobases (kb) in both the mouse and human genomes (WANG et al. 1984; SHTIVELMAN et al. 1986; BERNARDS et al. 1987, 1988). In the mouse it consists of four alternative 5′ exons, called Ia, Ib, II, and III, followed by 10 or 11 common exons (WANG et al. 1984; BEN-NERIAH et al. 1986a; BERNARDS et al. 1988). Two of the alternative first exons are used in virtually all murine cells to make two alternative forms of c-*abl* messenger mRNA, a type Ia (5.5 kb) or a type Ib (6.5 kb) molecule, by a process of alternative splicing (BEN-NERIAH et al. 1986a; BERNARDS et al. 1988). Although both mRNAs are found in somatic cells the level of the type Ib transcript appears constant whereas the level of the type Ia transcript varies over a 10-fold range (RENSHAW et al. 1988). The other two murine 5′ exons are also used to make RNA by alternative splicing but the frequency of these RNAs is very low (BEN-NERIAH et al. 1986a). A 4-kb testis-specific c-*abl* transcript has also been identified (MULLER et al. 1982; MEIJER et al. 1987; OPPI et al. 1987). This transcript is like the type Ia c-*abl* mRNA, but it is truncated for 1.2 kb at its 3′ end by termination of transcription

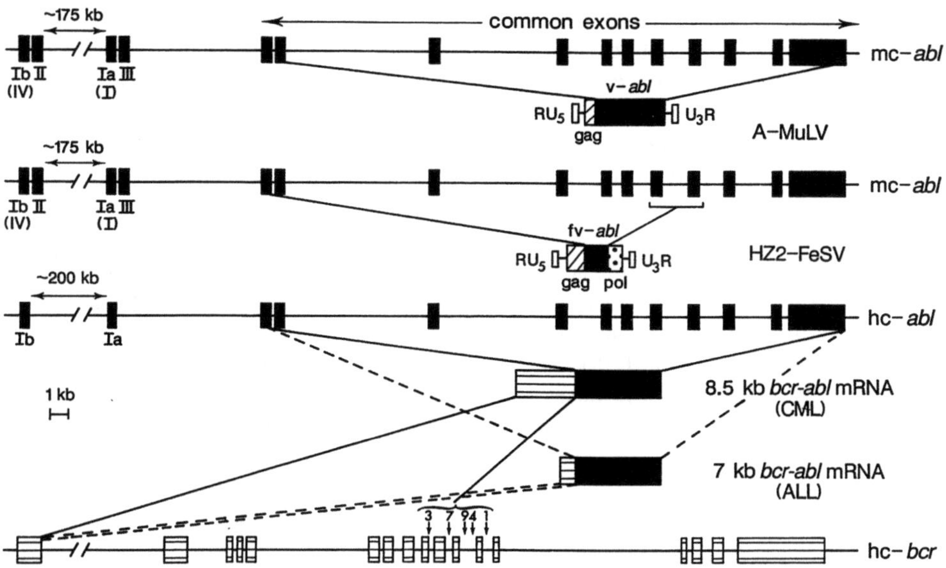

Fig. 1. The structures of *abl* proto-oncogenes and their activated derivatives. The murine c-*abl* proto-oncogene (mc-*abl*) is drawn approximately to scale and is presented in comparison to A-MuLV and HZ2-FeSV, although the c-*abl* sequences of the latter were presumably derived from the feline c-*abl* gene. The CML and ALL forms of the *bcr-abl* mRNAs are drawn in comparison to the human c-*abl* (hc-*abl*) and human *bcr* (hc-*bcr*) genes. *Black boxes* represent c-*abl* exon sequences and *striped boxes* represent *bcr* exon sequences. The *arrows* over the *bcr* gene indicate the approximate positions at which fusion of *bcr* and *abl* sequences have occured in CML, and the *numerals* indicate the number of times an indicated position has been used for fusion of *bcr* and *abl* sequences in CML. Boxes with diagonal stripes represent MuLV *gag* sequences, and the stippled box represents FeLV *pol* sequences.

(Oppi et al. 1987). Only exons Ia and Ib have been identified in the human, with corresponding mRNAs of 6 and 7 kb (Westin et al. 1982; Shtivelman et al. 1986). Exon Ib of the mouse or human lies approximately 200 kb 5′ of the first common exon (Bernards et al. 1987, 1988) and encodes a protein with a potential signal for myristylation, unlike the other alternative 5′ exons (Ben-Neriah et al. 1986a).

2.1 A-MuLV

The recombination event(s) that generated A-MuLV from M-MuLV and the c-*abl* proto-oncogene resulted in the deletion of 6285 bases from M-MuLV and the insertion of 3879 bases of processed c-*abl* sequences (Shinnick et al. 1981; Reddy et al. 1983; Lee et al. 1985). As a consequence of the deletion of sequences from M-MuLV, A-MuLV is incapable of replication in the absence of a helper virus (*vide infra*). The 5′ recombination event took place within the third exon of the c-*abl* proto-oncogene (Wang et al. 1984) and within the *gag* gene at a position encoding the first 21 amino acids of the CA molecule (p30) (Reddy et al. 1983) to generate an in-frame fusion of the *gag* and *abl* genes, the *gag*-v-*abl* gene. The 3′ recombination point in the processed c-*abl* gene occurred 856 bases 3′ to the c-*abl* protein termination codon (WANG et al. 1984; LEE et al. 1985; OPPI et al. 1987). The 3′ recombination point in M-MuLV interrupted the *env* gene, thereby disrupting the reading frame of that gene (REDDY et al. 1983) (Fig. 1). The generation of v-*abl* from c-*abl* may also involve mutation of c-*abl* sequences. One point mutation has been reported in the c-*abl*-derived sequences of the v-*abl* gene. However, its significance in the oncogenic activity of v-*abl* is not known (OPPI et al. 1987). Reexamination of the nucleotide sequence of v-*abl* indicates that this point mutation is the only difference between c-*abl* and v-*abl* (M. PASKIND and D. BALTIMORE, personal communication). There are two strains of A-MuLV in widespread use. The P120 strain encodes a 120-kDa *gag*-v-*abl* protein and was derived from the P160 strain by deletion of 789 bases from the v-*abl* gene. The P160 strain encodes a 160-kDa *gag*-v-*abl* protein (ROSENBERG and WITTE 1980; LATT et al. 1983) (Fig. 2).

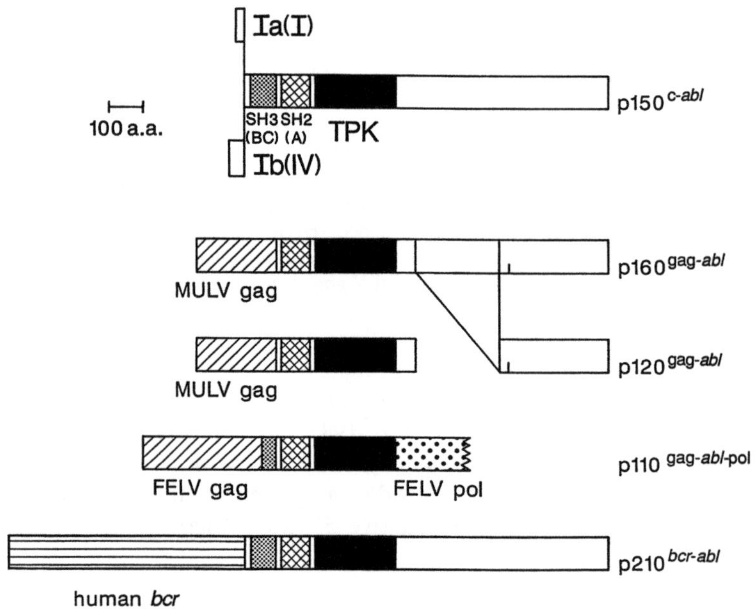

Fig. 2. The structures and homology regions of Abelson proteins. The regions of Abelson proteins homologous to pp60^{src} are the tyrosine protein kinase domain (*TPK, black*), the *src* homology region 2 SH2, *cross-hatched*), and the *src* homology region 3 (SH3, *dotted*). Retroviral *gag*-derived sequences are indicated by *diagonal lines* and *bcr*-derived sequences are indicated by horizontal lines

2.2 HZ2-FeSV

A feline isolate of an activated v-*abl* oncogene, fv-*abl*, has been recovered from a sarcoma that developed in a cat inoculated with feline leukemia virus (FeLV)

(Besmer et al. 1983). This virus, HZ2 feline sarcoma virus (HZ2-FeSV), when inoculated as a complex with FeLV, induces polyclonal sarcomas in newborn kittens (Bergold et al. 1987a). Nucleotide sequence analysis indicates that, like A-MuLV, HZ2-FeSV also encodes a *gag-abl* fusion protein. In HZ2-FeSV *gag-abl* sequences are also fused to *pol* sequences to encode a *gag*-fv-*abl-pol* fusion protein, $p110^{gag\text{-}fv\text{-}abl\text{-}pol}$ (Bergold et al. 1987b). Unlike A-MuLV, the HZ2-FeSV recombination event took place in the sequences which are homologous to those of the second exon of the murine c-*abl* gene (Bergold et al. 1987b). (The feline c-*abl* gene or complementary cDNAs have not been sequenced.) Thus, HZ2-FeSV contains additional sequences derived from the 5′ portion of the c-*abl* proto-oncogene as compared to A-MuLV. It also encodes 344 amino acids of *gag* information corresponding to the p15 and p10 mature proteins and 148 amino acids of the p30 mature protein (Bergold et al. 1987b). HZ2-FeSV lacks sequences derived from the 3′ exons of the c-*abl* proto-oncogene and encodes only 439 amino acids of *abl*-derived information (Bergold et al. 1987b). The 3′ breakpoint of *abl* sequences in HZ2-FeSV is very close to that identified in mutants of A-MuLV as the 3′ limit of sequences necessary for transformation of lymphoid cells in vitro (Prywes et al. 1983, 1985b; Bergold et al. 1987b). *pol*-derived sequences encode at least 200 additional amino acids at the 3′ terminus of the *gag*-fv-*abl-pol* fusion protein (Bergold et al. 1987b) (Fig. 1).

2.3 *bcr-abl*

Forms of activated *abl* oncogenes are also found in human leukemia. Virtually all examples of CML and 10%–20% of ALL contain chromosomal translocations (the Philadelphia chromosome, Ph′) (Nowell and Hungerford 1960; Rowley et al. 1973) that juxtapose *bcr* (*b*reakpoint *c*luster *r*egion) and c-*abl* sequences which are normally found on human chromosomes 22 and 9, respectively (Heisterkamp et al. 1983; Groffen et al. 1984; Hermans et al. 1987). Although the role of these rearranged genes is not known in CML or ALL, it seems likely that the recovery of these rearrangements in these two diseases is not fortuitous but rather reflects some selective growth advantage that *bcr-abl* genes confer on the leukemic clone. In CML the 5′ terminal five to eight exons and introns of the *bcr* gene are fused to sequences 5′ to the c-*abl* gene or sequences within the first very large c-*abl* intron (Heisterkamp et al. 1985; Hermans et al. 1987). The breakpoint within the *bcr* gene in CML occurs over a region spanning 5.8 kb and containing five *bcr* exons (Groffen et al. 1984; Heisterkamp et al. 1985; Shtivelman et al. 1985; Grosveld et al. 1986) (Fig. 1). Although there is heterogeneity in the breakpoints within the *bcr* and c-*abl* genes in different CML patients, a novel 8.5-kb *bcr-abl* mRNA has been found in the leukemic cells of all patients carrying the *bcr-abl* translocation (Canaani et al. 1984; Collins et al. 1984; Gale and Canaani 1984; Nen-Neriah et al. 1986a). This mRNA appears to join at least five *bcr* exons with the second c-*abl* exon (first common exon) by a process of alternative slicing (Ben-Neriah et al. 1986a; Shtivelman et al. 1986; Mes-Masson et al. 1986). Thus, although the alternative exon Ia is retained in the portion of the c-*abl* gene participating in *bcr-abl* translocations, this exon is removed from the *bcr-abl* mRNA (SHTIVELMAN et al. 1986) (Figure 1). The *bcr-abl* fusion gene encodes a product of 210-kDa, the *bcr-abl* fusion protein (KONOPKA et al. 1984; BEN-NERIAH et al. 1986b).

Although less well characterized, *bcr-abl* gene translocations are also found in ALL (HERMANS et al. 1987; RUBIN et al. 1988). In this case, however, only the 5′-most exon of the *bcr* gene participates in the translocation, and the translocation breakpoints in c-*abl* appear to take place at the same position as in CML (HERMANS et al. 1987). Once again, alternative splicing results in a hybrid transcript containing 5′ *bcr* and 3′ *abl* sequences. In ALL, however, a 7-kb mRNA is found (FAINSTEIN et al. 1987; HERMANS et al. 1987). The ALL *bcr-abl* mRNA contains the 5′-most *bcr* exon joined to the second *abl* exon, thereby eliminating any of the alternative 5′ c-*abl* exons. The 7-kb ALL-specific *bcr-abl* mRNA encodes a 190-kDa protein (KURZROCK et al. 1987; WALKER et al. 1987; CLARK et al. 1988). The basis for the striking difference between CML and ALL *bcr-abl* mRNAs is not known. It could reflect the time in hematopoietic development when particular *bcr* sequences are sensitive to chromosomal breakage. Alternatively, it could reflect different pathologic potentials associated with different forms of the resulting proteins. In the latter case selection on the basis of disease would play a major role in the association of particular rearrangements with CML or ALL.

3 The Abelson Proteins: Homologies to Other Viral Oncogenes and Biochemical Activities

3.1 Regions of Abelson Proteins Homologous to Other Proto-oncogene Products

The first evidence that the v-*abl* protein was a member of the family of tyrosine protein kinases (TPK) was the observation that the v-*abl* protein, like $pp60^{v\text{-}src}$, showed phosphorylating activity for tyrosine residues when the protein was immunoprecipitated and immune precipitates were incubated under appropriate reaction conditions (WITTE et al. 1980; SEFTON et al. 1981). More definitive evidence for this hypothesis came from the characterization of v-*abl* protein expression and TPK activity in *E. coli*, which lacks tyrosine kinases (WANG et al. 1982). Nucleotide sequence analysis of c-*abl* cDNAs, A-MuLV and HZ2-FeSV proviral DNAs, and *bcr-abl* cDNAs has revealed three principal areas of sequence similarity to other tyrosine protein kinases (GROFFEN et al. 1983; REDDY et al. 1983; BEN-NERIAH et al. 1986a; SADOWSKI et al. 1986) (Fig. 2). The Abelson proteins all contain a sequence of approximately 250 amino acids that is highly conserved among all members of the TPK family, and constitutes the catalytic domain of the enzyme. Hallmarks of this domain are the ATP-binding site containing a conserved lysine residue (Lys-392) (KAMPS et al. 1984) and the principal in vivo site for tyrosine phosphorylation at Tyr-514 (corresponding to v-*src* Tyr-416) (KLOETZER et al. 1985; KONOPKA and WITTE 1985). Further support for the functional similarity of the TPK domain of *abl* proteins and those of other TPK, particularly $pp60^{v\text{-}src}$, comes from experiments in which *src*-homologous residues of v-*abl* were mutated to amino acids corresponding to thermosensitive kinase mutants of pp60$^{v\text{-}src}$. The mutated *abl* proteins were thermosensitive for kinase activity, at least in the context of truncated v-*abl* proteins (ENGELMAN and ROSENBERG 1987).

Two additional regions of sequence similarity are shared between *abl* proteins and those members of the TPK family that lack transmembrane domains and are presumably localized in the cytoplasm or anchored in the plasma membrane. One domain, SH2 (SADOWSKI et al. 1986; DE CLUE et al. 1987), shared with v-*src*, v-*fps*, and other TPK enzymes, is retained in all activated *abl* derivatives. In the v-*fps* protein some mutations in this domain render the protein thermosensitive for TPK activity and host cell-dependent for transformation (SADOWSKI et al. 1986; DE CLUE et al. 1987). On this basis it has been hypothesized that the SH2 domain of TPK proteins modulates kinase activity and may determine the interaction of TPK with other regulatory proteins. Comparison of the c-*abl* sequence to c-*src* and many other TPKs (but not v-*fps* or c-*fps*) reveals an additional region of homology, called SH3, located 5′ to the SH2 domain.

The SH3 domain along with SH2 has been transduced from a gene resembling that for phospholipase C in the v-*crk* oncogene in the absence of other c-*rk* proto-oncogene sequences (MAYER et al. 1988; STAHL et al. 1988). The recovery of a transforming virus that contains only the SH2 and SH3 domains and no TPK domain raises the possibility that both SH2 and SH3 are regulatory moieties or encode an additional enzymatic activity not yet described for the TPK proteins.

Interestingly, the SH3 domain is deleted from A-MuLV, and truncated in HZ2-FeSV. Preliminary experiments suggest that deletion of sequences within this region may play a necessary role in the activation of c-*abl* proteins to transforming proteins (*vide infra*).

3.2 The Abelson Kinases

In view of the quite different amino-terminal sequences found in v-*abl*, *bcr-abl*, and c-*abl* proteins and the different biological activities of these proteins (*vide infra*), one might expect that the kinase activities of these proteins would be quantitatively or qualitatively different. Although initial reports indicated this was the case (PONTICELLI et al. 1982; KONOPKA et al. 1984; DAVIS et al. 1985), alterations in the in vitro conditions for kinase assay led to the recognition that all of these proteins can show quite respectable in vitro kinase activities (KONOPKA and WITTE 1985), and differ only about twofold in the K_m for ATP or V_{max} for ATP hydrolysis (PENDERGAST et al. 1987). The specific autokinase activity of the v-*abl* protein is approximately ten times that of the c-*abl* protein when both are expressed in COS cells (WANG 1988). It is also clear that expression of the activated v-*abl* or *bcr-abl* gene in mouse or human cells leads to a five- to tenfold elevation in the number of phosphotyrosine residues (SEFTON et al. 1981; KONOPKA et al. 1984). This elevation of phosphotyrosine residues requires the continued expression of a functional *abl* kinase (KIPREOS et al. 1987; ENGELMAN and ROSENBERG 1987). Data also indicates that elevated expression of the c-*abl* kinase does not lead to elevated in vivo levels of phosphotyrosine (WANG 1988). In vivo, Tyr-514 of v-*abl* is phosphorylated and other tyrosine residues of *bcr-abl* proteins are phosphorylated (Konopka et al. 1984; Konopka and Witte 1985). Phosphotyrosine residues have not been detected on c-*abl* proteins in vivo, whereas phosphoserine residues have been detected (Ponticelli et al. 1982; Konopka et al. 1984).

4 Functional Domains of the v-*abl* Protein and Activation of c-*abl* to v-*abl*

Before discussing the effects of mutations in A-MuLV on the biochemical and biological functions of the *gag*-v-*abl* protein, it may be useful to list the variety of in vitro assays used to demonstrate the biological effects of the virus (Table 1).

Table 1. Common assays for cell transformation by activated Abelson oncogenes

Cell	Assay	Reference
NIH3T3	Focus formation	Scher and Siegler 1974
Mouse bone marrow	Proliferation and cell line establishment	Rosenberg et al. 1975
Mouse bone marrow	Colony formation in semisolid agarose	Rosenberg and Baltimore 1976
IL-3-dependent hematopoietic cell lines	IL-3 independent proliferation	Pierce et al. 1985; Cook et al. 1985
Long-term feeder layer-dependent mouse bone marrow cultures	Enhanced proliferation and eventual feeder-layer independence	Whitlock et al. 1981

Perhaps least reflective of its tumor-inducing activity in mice is the ability of A-MuLV to cause morphological transformation of established fibroblastoid cell lines such as NIH3T3 (Scher and Siegler 1975). Although fibrosarcomas have never been observed in any mouse infected with A-MuLV (Risser (1982), the NIH3T3 transformation assay is the standard assay for A-MuLV infectivity because of its sensitivity and ease of quantitation. Two additional tissue culture assays for A-MuLV are based respectively on the ability of A-MuLV to induce proliferation of primary bone marrow cells in agarose suspension (Rosenberg and Baltimore 1976), and the ability of A-MuLV to induce factor-independent growth in a variety of cell lines dependent on the multilineage growth factor interleukin-3 (IL-3) for continued proliferation (Oliff et al. 1985; Pierce et al. 1985; Cook et al. 1985). In addition A-MuLV can stimulate the proliferation of bone marrow cells infected in mass culture and lead to their establishment as immortal cell lines (Rosenberg et al. 1975).

4.1 Functional Domains Identified by Deletion and Truncation Mutants

Mutants deleted for regions of the viral genome have been used to establish the minimum sequences of the *gag*-v-*abl* protein necessary for transformation of NIH3T3 cells (Fig. 3). Minimum transforming sequences identified to date include the amino terminus of *gag*, the SH2 domain and the TPK domain (PRYWES et al. 1983, 1985a; SRINIVASSAN et al. 1982). The amino-terminal *gag* sequences contain a signal for myristylation of the protein, which is necessary for localization of the protein at the plasma membrane (SEFTON et al. 1981; HENDERSON et al. 1983; BUSS and SEFTON 1985). The sequences encoded by the 5′ portion of the *gag* gene in A-MuLV can be substituted with sequences of v-*src* which also encode a myristylation signal, and the resulting hybrid gene retains its transforming activity for NIH3T3 cells and for bone marrow lymphocytes (MATHEY-PREVOT and BALTIMORE 1985). Deletion mutations which interrupt the TPK domain or which overlap the *gag*-SH2 junction abolish transforming activity for NIH3T3 cells (PRYWES et al. 1985a; SRINIVASSAN et al. 1982). Additional *gag* sequences appear to be

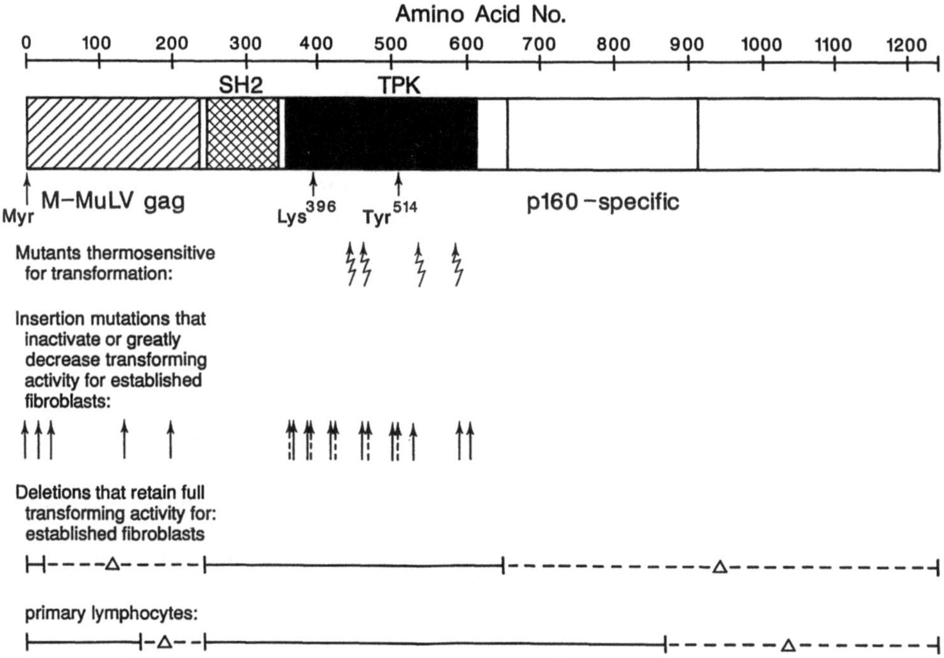

Fig. 3. Functional domains of the A-MuLV *gag*-v-*abl* protein. The site of myristylation (*Myr*), the ATP-binding site of the tyrosine protein kinase domain (*TPK*) (Lys-396), and the site of in vivo phosphorylation (Tyr-514) are indicated. The portion of *gag*-v-*abl* deleted in A-MuLV P120 strains is indicated (*p160-specific*). The positions of the insertion mutations in A-MuLV P160 that abrogate transforming activity are indicated by *solid arrows*, whereas *dashed arrows* indicate mutations in a truncated strain of A-MuLV. *Dashed horizontal lines* indicate sequences removed from deletion mutants

required to transformation of bone marrow lymphocytes in agarose suspension or mass culture (PRYWES et al. 1985b). The absence of these *gag* sequences from A-MuLV results in the production of *gag*-v-*abl* proteins that have very short half-lives in lymphocytes (PRYWES et al. 1985b).

A series of mutants which appear to prematurally terminate synthesis of the Abelson protein or are deleted for sequences located in the carboxyl half of the Abelson protein show reduced transforming activities in the agarose transformation assay and have less efficient in vitro kinase activities (ROSENBERG et al. 1980; GOFF et al. 1981; PRYWES et al. 1985a). Some mutant A-MuLVs which lack carboxyl portions of v-*abl* show decreased in vitro transforming potential yet retain full tumorigenic activity (MURTAGH et al. 1986). Thus, sequences in the carboxyl half of the Abelson protein can modulate the transforming activity of the protein. Other workers have indicated a role for this region in the toxicity of A-MuLV for certain established fibroblastoid cell line (WATANABE and WITTE 1983).

4.2 Functional Domains Defined by Insertion and Thermosensitive Mutants

Analysis of a series of insertion mutations for their effects on the transforming activity of A-MuLV for NIH3T3 cells has confirmed the picture derived from studies of deletion mutations with certain important exceptions. Both methods indicated the essential role of the TPK domain for transformation (PRYWES et al. 1985a; REES-JONES et al. 1988), but all insertion mutations made in the *gag* region reduced the transforming efficiency of the virus (REES-JONES et al. 1988). Many mutations made in the SH2 domain or in the carboxyl portion of the v-*abl* protein had no effect on transforming activity. The role of the SH2 domain, therefore, remains to be established for A-MuLV. One insertion mutation made in the P160-specific domain showed a partial transforming ability in NIH3T3 cells but was able to render the cells completely platelet-derived growth factor (PDGF) independent (REES-JONES et al. 1988).

Thermosensitive mutants of v-*abl* kinase have been isolated by screening directly for such mutants following insertional mutagenesis and expression in *E. coli* (KIPREOS et al. 1987), by site-directed mutagenesis of amino acids that occupy homologous positions in v-*src* and v-*abl* (ENGELMAN and ROSENBERG 1987), and by chemical mutagenesis and screening for temperature-sensitive differentiation phenotype (TAKEMORI et al. 1987). Molecular analyses of the mutations derived by the first two methods have confirmed that each lies within the TPK domain, and the mutant proteins have been shown to be thermosensitive for tyrosine kinase activity when the protein was expressed in *E. coli*. These mutants also have been used to establish the requirement for continued expression of kinase activity for the maintenance of cell transformation as judged by thermosensitive morphological transformation of NIH3T3 cells and continued proliferation of bone marrow-derived transformed lymphocytes (ENGELMAN and ROSENBERG 1987; KIPREOS et al. 1987). Some of these mutants have also been used to demonstrate that continued expression of v-*abl* kinase is necessary to maintain factor-independence following infection of IL-3-dependent cell lines with A-MuLV (KIPREOS and WANG 1988).

4.3 Activation of the c-*abl* Proto-oncogene to the v-*abl* Oncogene

Two groups have investigated the ability of retrovirus vectors encoding c-*abl* cDNAs to transform NIH3T3 cells and in some cases lymphocytes (BALTIMORE 1987; WANG 1988). The results of these experiments have established that expression of neither type Ia nor type Ib c-*abl* cDNAs from a retrovirus vector is sufficient to achieve transformation of NIH3T3 cells (WANG 1988; P. JACKSON and D. BALTIMORE, personal communication). Moreover, the specific autokinase activities of c-*abl* type Ia or type Ib proteins was approximately tenfold less than that of the v-*abl* protein when assayed in immunoprecipitates from *E. coli* or monkey COS-1 cells (WANG 1988). Deletion of sequences in the first c-*abl* variable exons resulted in a three- to fivefold increase in the specific kinase activity of c-*abl* protein (WANG 1988). This deletion was not sufficient to achieve transformation of NIH3T3 cells in the absence of a myristylation signal from *gag* or from the amino terminus of the type Ib c-*abl* exon (FRANZ et al. 1988). Internal deletion of sequences from the first common exon of type Ib c-*abl* cDNA, or deletion of sequences from the first c-*abl* exon plus amino-terminal addition of *gag* sequences, when placed in a M-MuLV vector yielded viruses that transformed lymphocytes and induced tumors in mice (P. JACKSON and D. BALTIMORE, personal communication). Additional alterations in the amino terminus of c-*abl* or in sequences carboxyl to the TPK domain may be necessary to achieve the same efficiency of transformation of NIH3T3 cells or lymphocytes as is seen with v-*abl*, however (OPPI et al. 1987; WANG 1988; P. JACKSON and D. BALTIMORE, personal communication).

5 Transformation of Hematopoietic Cells In Vitro by Activated *abl* Oncogenes

The principal assays that have been used to assess the transforming effects of *abl* oncogenes utilize bone marrow or fetal liver cells as targets because these cells appear to be the most common in vivo targets for tumor induction. An additional biological assay for the effects of *abl* oncogenes is based on the observation that A-MuLV induces factor-independent growth in several lines of IL-3-dependent cells (Table 1). Recent evidence from several laboratories suggests that recovery of transformed lines from these assays is a complicated process that reflects not only the genotype of the input virus, as has been emphasized thus far in this review, but also the dynamics and selective conditions of the particular assay. These topics will now be addressed.

5.1 Transformation of Lymphoid Cells

Transformation of lymphoid precursor cells by A-MuLV has been quantified by infecting bone marrow cells, culturing the cells for a month or more in medium containing 2-mercaptoethanol, and scoring the frequency of cultures that give rise to established cell lines (Rosenberg et al. 1975). Under these conditions A-MuLV infection gives rise to a high frequency of established cell lines. Due to the continuous replication of virus in such cultures it has not been possible to determine the efficiency with which a given infected cell gives rise to an established cell line, or the number of cells in the culture that are capable of giving rise to established cell lines. When fetal liver cells are used as the target cells the resultant transformants are at the earliest stage of B-cell development and have just initiated rearrangement of the immunoglobulin heavy chain locus (Alt et al. 1981). Use of bone marrow cell targets from 3- to 5-week-old mice tends to yield transformed lines of a more differentiated phenotype characterized by expression of cytoplasmic μ chain and initiation of light chain gene rearrangement (Siden et al. 1979; Alt et al. 1981; Siden et al. 1981). A-MuLV infection of bone marrow cell cultures in the presence of the B-cell mitogen lipopolysaccharide (LPS) has yielded established cell lines with characteristics of B-cells frozen at a stage of differentiation immediately prior to antigen stimulation (Serunian and Rosenberg 1986). The established cell lines recovered from these experiments are invariably tumorigenic and usually express the *gag*-v-*abl* oncogene product (Rosenberg et al. 1975; Serunian and Rosenberg 1986).

A variation on the mass culture assay is the use of bone marrow cell cultures dependent on long-term bone marrow feeder layers (Whitlock-Witte feeder layers) for continued proliferation (Whitlock and Witte 1981). When such cultures were infected with A-MuLV the cell population initially grew poorly in the absence of the feeder layer and did not plate efficiently in agarose suspension in the absence of feeder layers. Culture for an additional 3 months resulted in the emergence of feeder-layer-independent cells (Whitlock and Witte 1981). Experiments with cell clones recovered shortly after culture and subcultured for an additional 3 months demonstrated that feeder layer independence could arise within a single clone of cells without alteration in the expression of the v-*abl* oncogene. The oncogenicity of the late passage clones was increased over that of the early passage clones (Whitlock et al. 1983). A *bcr-abl* gene expressed from a M-MuLV vector causes enhanced proliferation of feeder-layer-dependent lymphoid cells but does not cause morphological transformation of NIH3T3 cells (McLaughlin et al. 1987; Daley et al. 1987).

The recovery of feeder-layer-dependent A-MuLV-transformed cells implies that a continuum of transformed phenotypes can be recovered following A-MuLV infection, a result also found with transformation of fibroblastoid cells by DNA tumor viruses (Risser and Pollack 1974). These experiments also raise the possibility that tumorigenesis by A-MuLV requires events in addition to expression of the v-*abl* oncogene.

Quantitation of A-MuLV transformation has been greatly improved by the modification of the mass culture assay to include semisolid agarose (Rosenberg and Baltimore 1976). Bone marrow or fetal liver cells infected with A-MuLV are plated at a density of 5×10^5 cells/ml in semisolid agarose, and macroscopic colonies containing approximately 10^5 cells are scored 9–14 days later. The number of colonies is linearly dependent on the dose of virus (Rosenberg and Baltimore 1976), but the number of colonies per fibroblast focus-forming unit (FFU) varies considerably, i.e. from one lymphoid colony per 10^3 FFU to one lymphoid colony per 10^5 FFU (ROSENBERG and BALTIMORE 1976; WITTE et al. 1980; GREEN et al. 1987a, b; PALUMBO et al. 1988; TIDMARSH et al. 1988). An additional important consideration is that the number of transformed colonies detected is not linearity dependent on the number of cells plated but is influenced by cell density (PALUMBO et al. 1988; TIDMARSH et al. 1988). One group has demonstrated that supplementation of the medium with supernatants from long-term bone marrow feeder layers and an A-MuLV-transformed NIH3T3 cell line can increase the number of colonies three- to fourfold (PALUMBO et al. 1988). Taken together these results could be interpreted to mean that the agarose colony assay is the most selective system for A-MuLV transformation. Infection of mouse bone marrow with viruses expressing the feline v-*abl* gene yields transformants in this assay (N. ROSENBERG and P. BESMER, personal communication).

Using cell fractionation techniques it has been possible to identify a population of cells in bone marrow that is 10- to 20-fold enriched for A-MuLV target cells detected in the agarose assay (TIDMARSH et al. 1988). These cells constitute 1%–2% of total bone marrow and express the B-lymphoid lineage marker B220 as well as low densities of the Thy-1 cell surface antigen which is found on stem cells and all T cells in the hematopoietic system. This population is also enriched for precursor cells that can generate B cells following in vivo reconstitution of lethally irradiated mice.

5.2 Transformation of Factor-Dependent Cell Lines

Abrogation of growth factor dependence following A-MuLV infection was first observed with myeloid or mast cell lines that had been chronically infected with the virus (OLIFF et al. 1985; PIERCE et al. 1985; COOK et al. 1985). Under these conditions cell lines were obtained which expressed the Abelson protein, were IL-3 independent for growth and did not produce IL-3- or IL-3-specific mRNA. Such results indicated that expression of v-*abl* could transform cells to factor independence by a nonautocrine mechanism.

Experiments with thermosensitive mutants established that continued proliferation in the absence of factor required continued expression of a functional v-*abl* kinase (KIPREOS and WANG 1988). Using a myeloid line that can differentiate in culture in response to granulocyte-macrophage colony-stimulating factor (GM-CSF), ROVERA and colleagues (1987) were able to demonstrate that expression of the v-*abl* oncogene blocked differentiation of these cells. Comparison of the activity of other tyrosine kinase oncogenes to v-*abl* suggests that v-*abl* is considerably more efficient than v-*src*, v-*fms*, or v-*fps* in inducing IL-3 independence in lymphoid and myeloid cell lines (MATHEY-PREVOT and BALTIMORE 1986; MCCUBREY et al. 1988). The establishment of IL-3 independence following infection of lymphoid lines with A-MuLV results in additional changes in the expression of genes such as those for the IL-2 receptor and MHC class I antigens (MCCUBREY et al. 1989). Quantitative virological studies of lymphoid and myeloid factor-dependent lines indicate that only a minority of the cells infected with A-MuLV become factor independent immediately after infection (MCCUBREY et al. 1989; VOGT et al. 1988) and that in some cases factor independence arises concomitant with the appearance of chromosomal aberrations in the cell lines (VOGT et al. 1988). The *bcr-abl* oncogene is also active in this assay system, although with lower efficiency than the *gag*-v-*abl* oncogene (G. Q. DALEY and D. BALTIMORE, personal communication).

6 Tumor Induction by Activated *abl* Oncogenes

Central problems in understanding the pathogenesis of diseases triggered by *abl* oncogene expression include (a) the biological and molecular bases of the selective association of *abl* oncogenes with particular hematopoietic diseases, and (b) the role, if any, of secondary events in tumor progression. Potentially related to these questions is the role of helper virus in those cases where disease is induced by a complex of helper and defective viruses.

6.1 Determinants of Tissue Tropism

The large majority of mice infected intravenously or intraperitoneally with A-MuLV develop tumors of pre-B cell phenotype (PREMKUMAR et al. 1975; SKLAR et al. 1975; GREEN et al. 1987a, b, 1988a) approximately 3–4 weeks after infection. The incidence of tumors is linearly dependent on the dose of virus, and the sensitivity of mice to tumor induction is affected by both the age and the genotype of the mouse (ABELSON and RABSTEIN 1970b; RISSER et al. 1978). Infection of mice that had been inoculated previously with the adjuvant pristane resulted in the rapid appearance of plasmacytomas (POTTER et al. 1973). Thus, both early and later stages of the B-lymphoid lineage are susceptible to A-MuLV tumor induction.

A high incidence of thymic lymphomas has been obtained following intrathymic inoculation of some strains of mice with A-MuLV (COOK 1982; RISSER et al. 1985). Characterization of the cell surface phenotypes, and rearrangements of T-cell receptor and immunoglogulin genes in these tumors revealed that the tumors were of heterogeneous phenotype. Some A-MuLV-induced thymic lymphomas corresponded to well defined T-cell subsets, whereas others had characteristics one might expect for pre-T cells (SCOTT et al. 1986; COOK and BALATON 1987; SPOLSKI et al. 1988). Helper virus-free A-MuLV efficiently induces thymic lymphomas (R. RISSER, unpublished data), and therefore it is likely that thymic lymphoma induction by A-MuLV is a direct consequence of infection of target cells in the thymus rather than infection of thymic epithelium. Some A-MuLV-induced thymic tumors are in the process of rearranging γ and δ T-cell receptor loci (G. HOLLAND, K. ITO, D. KAEHLER, S. TONEGAWA, and R. RISSER, unpublished data). One study reported the recovery of myeloid lines following in vitro cultivation of tumor material (RASCHKE et al. 1978), and mast cell lines have also been recovered from A-MuLV-infected mice (PIERCE et al. 1985). The frequency of myeloid or mast cell tumors appears to be very low, however. Thus, in mice the preferred targets for tumorigenesis by A-MuLV appear to be cells in early stages of B- or T-cell development, and the final outcome may depend on the microenvironment in which the transformed cell originates, e.g., thymus vs bone marrow.

In contrast to the murine lymphoid diseases induced by A-MuLV, *bcr-abl* translocations that generate a 8.5-kb fused transcript are associated with a myeloid or stem-cell neoplasia in humans. About 25% of CML cases develop B- or T-cell lymphoid blast crisis characterized by the presence of the Philadelphia chromosome (Ph') (BAKHSHI et al. 1983; GRIFFEN et al. 1983; CHAMPLIN and GOLDE 1985). Even in these cases the myeloid lineage is marked by Ph', and the *bcr-abl* fusion gene is expressed in myeloid cells (CHAMPLIN and GOLDE 1985). The minor fraction of ALL that contain *bcr-abl* translocations do not appear to form a special subset of ALL (CLARK et al. 1988). Thus, in humans activated *abl* genes are most commonly found in myeloid or hematopoietic stem cell tumors and to a lesser extent in tumors which appear to be of lymphoid origin. In contrast, the activated *abl* gene of cats was recovered from a fibrosarcoma, and extracts of that tumor induce fibrosarcomas with a long latent period in cats (BERGOLD et al. 1987a).

How can these diverse patterns be resolved? Is there a relationship between the particular form of the activated *abl* gene and the disease it instigates, or is the disease type dictated by the viral vector or state of the cell in which the translocation takes place? The answer to this question is by no means clear. To determine if the viral vector dictated the cell type of the tumor, GREEN et al. (1988a) substituted the U3 sequences of the long terminal repeat (LTR) of A-MuLV with those of the myeloerythrotropic Friend MuLV or the fibroerythrotropic Harvey Murine Sarcoma virus (MSV). (In other systems U3 sequences have been shown to be the major determinant of disease specifity; CHATIS et al. 1983.) These substitutions did not alter the pattern of disease induced by v-*abl*, although the Friend MuLV substitution did abrogate in vitro transforming ability (GREEN et al. 1988a). From these results one might conclude that the vector does not determine the disease tropism of A-MuLV, but rather that the viral oncogene is the determinative factor. Work with v-*src*/v-*abl* recombinant genes in a M-MuLV vector supports this conclusion (MATHEY-PREVOT et al. 1988).

If the *abl* oncogene determines the type of the disease then one might expect that transfer of fv-*abl* or *bcr-abl* genes to M-MuLV vectors would result in fibrosarcomas or myeloid leukemia upon inoculation into mice. Transfer of the fv-*abl* oncogene to a M-MuLV vector resulted in a virus that exclusively induced pre-B-cell tumors in mice (BERGOLD et al. 1987b), and transfer of the *bcr-abl* gene to a M-MuLV vector resulted in a virus that induces pre-B-cell proliferation in vitro (MCLAUGHLIN et al. 1987). To further complicate this issue, mice transgenic for the *gag*-v-*abl* or *bcr-abl* genes under the control of the immunoglobulin heavy chain enhancer and the SV40 promoter develop predominantly T-cell tumors rather than B-cell tumors (HARRIS et al. 1988). This pattern of disease may relate to the resistance of mice of that genetic background to A-MuLV pre-B-cell lymphomagenesis and their sensitivity to T-cell lymphomagenesis (RISSER et al. 1978; COOK 1982; RISSER et al. 1985).

One hypothesis that may accommodate the data on *abl* tissue tropism is that activated *abl* genes have no absolute lineage preference aside from very primitive cells of the hematopoietic system. The final expression of disease may reflect the dynamics of the cell population containing the oncogene in a particular species or organism. Some support for the idea that A-MuLV targets are primitive hematopoietic cells comes from experiments that involve infection of c-*myc* transgenic mice with other oncogenic viruses. Unlike viruses that encode v-*ras* or v-*raf* oncogenes, A-MuLV will not collaborate with the c-*myc* transgene to enhance transformation or tumorigenesis, perhaps because v-*abl* acts in a cell that has not yet transcriptionally activated the c-*myc* transgene (DYALL-SMITH and CORY 1988).

6.2 Role of Helper Virus in Disease Induction

A-MuLV-induced tumors are oligoclonal in origin as judged by provirus integration (SAVARD et al. 1987; GREEN et al. 1987b), although polyclonal infection of total bone marrow cells can be demonstrated by the midpoint of tumor latency (GREEN et al. 1987b). These results suggested that infection of the target cell and/or a subsequent process of clonal selection among infected cells is the rate-limiting step in the appearance of A-MuLV-induced tumors. Evidence supporting the second hypothesis has now been obtained in two different systems.

In one study Jolicoeur and colleagues have found that A-MuLV pseudotyped with non-leukemogenic helper viruses does not induce tumors in some strains of mice whereas A-MuLV pseudotyped with leukemogenic helper viruses does induce tumors in those mice (SAVARD et al. 1987), a result which confirms earlier studies (ROSENBERG and BALTIMORE 1978; SCHER 1978). By making recombinants of the helper viruses SAVARD et al. (1987) mapped the determinant for the helper effect in A-MuLV disease to the U3 sequences of the helper virus. Therefore, it seems likely that virus replication is the critical determinant of the helper virus effect because the U3 region acts in *cis* during virus replication and does not encode any viral proteins.

One possible mechanism for the involvement of helper virus replication is by insertional mutagenesis of a gene involved in additional steps needed for A-MuLV tumor induction. To identify such a gene, cellular sequences adjacent to sites of helper virus integration were molecularly cloned and used to screen a battery of A-MuLV tumor DNAs (POIRIER et al. 1988). A locus *Ahi-1* was found to be the site of helper virus integration in 16% of A-MuLV tumors, presumably because disruption of the normal function of *Ahi-1* provides the cell with a proliferative advantage. The locus has been mapped to mouse chromosome 10 and covers a region of 53 kb. The locus only rarely rearranged in other types of viral tumors. These results provide convincing evidence for participation of *Ahi-1* in the generation of some A-MuLV tumors, and suggest that tumorigenesis may involve multiple steps. Further support for a multistep mechanism of disease induction comes from studies of plasmacytomas induced by A-MuLV infection of pristane-treated mice. Both A-MuLV proviruses and rearranged c-*myc* genes are expressed in these rapidly developing tumors (OHNO et al. 1984).

6.3 Clonal Selection During Tumorigenesis

In a different study, Green and colleagues used mice infected with helper virus-free A-MuLV to directly identify preleukemic A-MuLV-infected cells and demonstrate their progression to leukemic cells as a consequence of clonal selection (GREEN et al. 1989). They found that weanling BALB/c mice developed a high frequency of monoclonal pre-B-cell tumors following infection with helper virus-free A-MuLV, and that the tumors did not express any contaminating helper virus (GREEN et al. 1987b). Although these results appear to contradict the results of SAVARD et al. (1987), different host mice, different routes of injection, and different virus coat proteins make comparisons of the two systems difficult.

Following infection of mice with helper virus-free A-MuLV, transformed cells were isolated from the mice at the midpoint of tumor latency and when tumors developed (Fig. 4). The biological and molecular properties of the preleukeic and leukemic cells were compared (Fig. 4; GREEN et al. 1989). Less than 5% of preleukemic cells had the capacity to give rise to tumors, whereas more than 85% of leukemic cells had the capacity to give rise to tumors. Comparison of the proviral integration patterns of preleukemic cells and leukemic cells from the same infected mouse indicated that both preleukemic and leukemic cell populations were members of the same infected clone. Comparison of the levels of v-*abl* autokinase activity in preleukemic and leukemic clones indicated that the levels of detectable

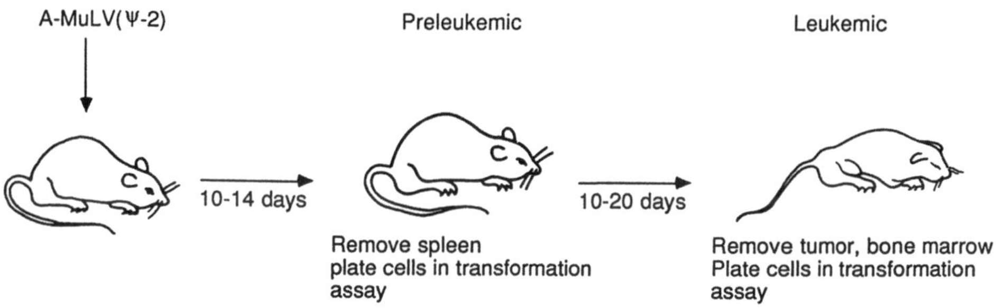

	Preleukemic	Leukemic
Transformed colonies/10^6 cells	10-50	300-1000
A-MuLV integration in colony DNA	Monoclonal	Monoclonal, same integration as preleukemic
gag-v-abl protein expression	1.0 ± 0.3	1.2 ± 0.3
Percent of colonies that are tumorigenic	3	85
Clonality of J_H rearrangements	Oligo- or polyclonal	Monoclonal

Fig. 4. Evidence for multiple events in the induction of A-MuLV tumors. Helper virus-free A-MuLV was used to infect mice, and the properties of individual preleukemic and leukemic colonies derived from these mice were compared following recovery from the agarose transformation assay. The level of *gag*-v-*abl* protein expression was quantified by the in vitro autokinase activity and has been normalized to that found in preleukemic cells (1.0). The percent of colonies that were tumorigenic was determined following subcutaneous inoculation of 3×10^4 cells into newborn mice

enzyme did not differ significantly. Within a given virus infected clone, multiple patterns of immunoglobulin heavy chain gene rearrangements were detected in the preleukemic population, whereas a single pattern was detected in the leukemic population. These results were interpreted to indicate that during the preleukemic phase of disease multiple subclones exist, and that by the leukemic phase of disease a particular clone has been selected on the basis of some trait other than the level of v-*abl* kinase activity. Thus, although the systems of POIRIER et al. (1988) and of GREEN et al. (1987b) differ, the central conclusion of both studies is that events in addition to v-*abl* expression are necessary for tumor formation.

Attempts have been made to relate the clonal selection observed in vivo to that documented in vitro (WHITLOCK et al. 1983; GREEN et al. 1987a, 1989). To determine if the degree of in vitro feeder layer independence correlated with the leukemic potential, the plating efficiencies of individual preleukemic or leukemic cells were determined in the presence or absence of long-term mouse bone marrow feeder layers. Although both populations showed very heterogeneous plating efficiencies with or without feeder layers, the plating efficiency of preleukemic cells was stimulated an average of 15-fold by the feeder layers while that of leukemic cells was not significantly stimulated (GREEN et al. 1989). However, a significant proportion of helper virus-free or helper virus-positive A-MuLV tumors do not grow in the presence of a mouse bone marrow feeder layer (GREEN et al. 1987b; D. KAEHLER and R. RISSER, unpublished observations). These results indicate that although there is a correlation of feeder layer independence with leukemic potential, the relation is by no means absolute, and feeder layer independence cannot be used as an indicator of tumorigenic potential, particularly when dealing with non-established cell populations.

A second system in which progression must play a central role in *abl*-initiated leukemogenesis is that of CML. CML patients in the chronic phase of the disease have an increased number of myeloid cells marked by the Ph′ chromosome (ROWLEY et al. 1973). This phase of disease can persist in a controlled state for several years (CHAMPLIN and GOLDE 1985). Moreover, during this more or less benign phase of the disease both lymphoid and myeloid cells are frequently marked by Ph′ (CHAMPLIN and GOLDE 1985). For reasons unknown, the disease then enters a blast crisis phase in which there is uncontrolled proliferation of Ph′-positive myelocytes or less frequently lymphocytes. Comparison of the levels of *bcr-abl* protein in chronic and blast phases of CML demonstrates that although *bcr-abl* protein levels may be elevated in the blast phase they are not always elevated (COLLINS et al. 1988). Thus, although it seems likely that the generation of the *bcr-abl* gene is necessary for the development of CML, it seems equally likely that the expression of the *bcr-abl* gene is not sufficient for the blast phase of disease. Defining the nature of these secondary steps in A-MuLV-induced disease and in CML presents a formidable challenge.

7 Summary and Future Directions

A major unresolved question for *abl*, as well as most other tyrosine kinase-encoding oncogenes, is the identity of the substrates critical to cell transformation.

Resolution of this question may need to await further experiments on the function of the normal c-*abl* proteins in cells of hematopoietic lineage. The transforming effects of activated *abl* proteins may reflect either the increased activity of activated *abl* kinases and consequent increase in the phosphorylated form of some critical cellular substrate(s) and/or the phosphorylation of substrates not normally phosphorylated. In this regard parallel quantitative comparisons of the kinase activities of various forms of activated *abl* proteins, e.g., fv-*abl*, v-*abl*, *bcr-abl*, and modified c-*abl*, with activities in a variety of biological assays could prove informative. Related to the question of substrate specificity is the question of normal regulation of c-*abl* kinase function. A key insight in this area may have been provided by the host range mutants of v-*fps* (SADOWSKI et al. 1986), and the discovery of the v-*crk* oncogene (MAYER et al. 1988). However, it remains to be determined to what extent c-*abl* protein function is regulated by sequences in its amino terminus and/or sequences located in the carboxyl portion of the protein.

Another unresolved question is the apparent tissue specificity of *abl* transformation and tumorigenesis. It seems likely that in the case of v-*abl*, critical determinants for tissue specificity reside in the viral oncogene (MATHEY-PREVOT 1987; GREEN et al. 1988a). How such determinants work, however, is totally unclear. It is also unclear to what extent *bcr* sequences contribute to the action of *bcr-abl* proteins in CML, and how and why the disease associated with the expression of *bcr-abl* proteins is different from v-*abl*-induced disease. Finally, the nature of the secondary events necessary for a complete malignant phenotype in cells expressing v-*abl* or *bcr-abl* is unknown. The identification of the *Ahi-1* locus and the isolation of preleukemic cells in v-*abl*-induced disease will hopefully provide insight into this question.

Acknowledgements. The authors thank Jean Y. J. Wang, F. Michael Hoffmann, and Patrick Green for their critical reading of the manuscript, and L. Michelle Bennett, Donal Kaehler, Eric Freed, Thomas Mitchell, and David Largaespada for patience and suggestions. Experiments from the authors' laboratory were supported by PHS grant CA 41302.

8 References

Abelson HT, Rabstein LS (1970a) Influence of prednisolone on Moloney leukemogenic virus in BALB/c mice. Cancer Res 30: 2208–2212

Abelson HT, Rabstein LS (1970b) Lymphosarcoma: virus-induced thymic-independent disease in mice. Cancer Res 30: 2213–2222

Alt F, Rosenberg N, Lewis S, Thomas E, Baltimore D (1981) Organization and reorganization of immunoglobulin genes in A-MuLV-transformed cells: rearrangement of heavy but not light chain genes. Cell 27: 381–390

Bakhshi A, Minowada J, Arnold A, Cossman J, Jensen JP, Whang-Peng J, Waldmann TA, and Korsmeyer SJ (1983). Lymphoid blast crisis in chronic myelogenous leukemia represent stages in development of B-cell precursors. N Engl J Med 309: 826–831

Baltimore D (1987) In: Melchers F, Potter M (eds) Mechanisms of B cell Neoplasia, Workshop at the Basel Institute for Immunology. Editiones Roche, Basel, pp 405–411

Ben-Neriah Y, Bernards A, Paskind M, Daley GQ, and Baltimore D (1986a). Alternative 5′ exons in c-*abl* RNA. Cell 28: 577–586

Ben-Neriah Y, Daley GQ, Mes-Masson A-M, Witte ON, and Baltimore D (1986b) The chronic myelogenous leukemia-specific P210 protein is the product of the *bcr/abl* hybrid gene. Science 235: 212–214

Bergold PJ, Wang JY-J, Hardy WD, Litlau V, Johnson E, and Besmer P (1987a) Structure and origin of HZ2-feline sarcoma virus. Virology 158: 320–329

Bergold PJ, Blumenthal JA, Andrea ED, Snyder HW, Lederman L, Silverstone A, Nguyen H, and Besmer P (1987b) Nucleic acid sequence and oncogenic properties of the HZ2 feline sarcoma virus v-*abl* insert. J Virol 61: 1193–1202

Bernards A, Rubin CM, Westbrook CA, Paskind M, and Baltimore D (1987) The first intron in the human c-*abl* gene is at least 200 kilobases long and is a target for translocation in chronic myelogenous leukemia. Mol Cell Biol 7: 3231–3236

Bernards A, Paskind M, and Baltimore D (1988) Four murine c-*abl* mRNAs arise by usage of two transcriptional promoters and alternative splicing. Oncogene 2: 297–304

Besmer P, Hardy WD, Zuckerman EE, Bergold P, Lederman L, Snyder JrHW (1983) The Hardy-Zuckerman 2-FeSV, a new feline retrovirus with oncogene homology to Abelson-MuLV. Nature 303: 825–828

Buss JE, Sefton BM (1985) Myristic acid, a rare fatty acid, is the lipid attached to the transforming protein of Rous sarcoma virus and its cellular homolog. J Virol 53: 7–12

Canaani E, Gale RP, Steiner-Saltz D, Aghai E, Berrebi A, Januszewicz A (1984) Altered transcription of an oncogene in chronic myeloid leukemia. Lancet 1: 593–595

Champlin RE, Golde DW (1985) Chronic myelogenous leukemia: recent advances. Blood 65: 1039–1047

Chatis PA, Holland CA, Hartley JW, Rowe WP, Hopkins N (1983) Role for the 3′ end of the genome in determining disease specificity of Friend and Moloney murine leukemia viruses. Proc Natl Acad Sci USA 80: 4408–4411

Clark SS, McLaughlin J, Timmons M, Pendergast AM, Ben-Neriah Y, Dow LW, Crist W, Rovera G, Smith SD, Witte ON (1988) Expression of a distinctive BCR-ABL oncogene in Ph′-positive acute lymphocytic leukemia (ALL). Science 239: 775–777

Collins SJ, Kubonishi I, Miyoshi I, Groudine MT (1984) Altered transcription of the c-*abl* oncogene in K-562 and other chronic myelogenous leukemia cells. Science 225: 72–74

Cook WD (1982) Rapid thymomas induced by Abelson murine leukemia virus. Proc Natl Acad Sci USA 79: 2917–2921

Cook WD, Balaton AM (1987) T-cell receptor and immunoglobulin genes are rearranged together in Abelson virus-transformed pre-B and pre-T cells. Mol Cell Biol 7: 266–272

Cook WD, Metcalf D, Nicola N, Burgess AW, Walker F (1985) Malignant transformation of a growth factor-dependent myeloid cell line by Abelson virus without evidence of a autocrine mechanism. Cell 41: 677–683

Daley GQ, McLaughlin J, Witte ON, Baltimore D (1987) The CML-specific P210 *bcr/abl* protein, unlike v-*abl*, does not transform NIH/3T3 fibroblasts. Science 237: 532–535

Davis RL, Konopka JB, Witte ON (1985) Activation of the c-*abl* oncogene by viral transduction or chromosomal translocation generates altered c-*abl* proteins with similar in vitro kinase properties. Mol Cell Biol 5: 204–213

de Clue JE, Sadowski I, Martin GS, Pawson T (1987) A conserved domain regulates interactions of the v-*fps* protein tyrosine kinase with the host cell. Proc Natl Acad Sci USA 84: 9064–9068

deKlein A, van Kessel AG, Grosveld G, Bartram C, Hagemeijer A, Bootsma D, Heisterkamp N, Groffen J, Stephenson JR (1982) A cellular oncogene is translocated to the Philadelphia chromosome in chronic myelocytic leukaemia. Nature 300: 765–767

Dyall-Smith D, Cory S (1988) Transformation of bone marrow cells from Eu-*myc* transgenic mice by Abelson murine leukemia virus and Harvey murine sarcoma virus. Oncogene Res 2: 403–409

Engelman A, Rosenberg N (1987) Isolation of temperature-sensitive Abelson virus mutants by site-directed mutagenesis. Proc Natl Acad Sci USA 84: 8021–8025

Fainstein E, Marcelle C, Rosner A, Canaani E, Gale RP, Dreazen O, Smith SD, Croce CM (1987) A new fused transcript in Philadelphia chromosome positive acute lymphocytic leukemia. Nature 330: 386–388

Franz WM, Berger P, Wang JY-J (1989) Deletion of an N-terminal regulatory domain of the c-*abl* tyrosine kinase activates its oncogenic potential. EMBO J 8: 137–148

Gale RP, Canaani E (1984) An 8-kilobase *abl* RNA transcript in chronic myelogenous leukemia. Proc Natl Acad Sci USA 81: 5648–5652

Goddard JM, Weilard JJ, Capeichi MR (1986) Isolation and characterization of *Caenorhabditis elegans* DNA sequence homologous to the v-*abl* oncogene. Proc Natl Acad Sci USA 83: 2172–2176

Goff S, Gilboa E, Witte ON, Baltimore D (1980) Structure of the Abelson murine leukemia virus genome and the homologous cellular gene: studies with cloned viral DNA. Cell 22: 777–785

Goff SP, Witte ON, Gilboa E, Rosenberg N, Baltimore D (1981) Genome structure of Abelson murine leukemia virus variants: proviruses in fibroblasts and lymphoid cells. J Virol 38: 460–468

Goff SP, Tabin CJ, Wang JY-J, Weinberg R, Baltimore D (1982) Transfection of fibroblasts by cloned Abelson murine leukemia virus DNA and recovery of transmissible virus by recombination with helper virus. J Virol 41: 271–285

Green PL, Kaehler DA, Risser R (1987a) Clonal dominance and progression in Abelson murine leukemia virus lymphomagenesis. J Virol 61: 2192–2197

Green PL, Kaehler DA, Risser R (1987b) Cell transformation and tumor induction by Abelson murine leukemia virus in the absence of helper virus. Proc Natl Acad Sci USA 84: 5932–5936

Green PL, Kaehler DA, McKearn JP, Risser R (1988a) Substitution of the LTR of Abelson murine leukemia virus does not alter the cell type of viral-induced tumors. Oncogene 2: 585–592

Green PL, Kaehler DA. Bennett LM, Risser R (1989) Multiple steps are required for the induction of tumors by Abelson murine leukemia virus. J Virol 63: 1989–1994

Griffen JD, Tantravahi R, Canellos G, Wisch JS, Reinherz EL, Sherwood G, Beveridge RP, Daley JF, Lane H, Schlossman SF (1983) T cell surface antigens in a patient with blast crisis of chronic myelogenous leukemia. Blood 61: 640–644

Groffen J, Heisterkamp N, Reynolds FH, Stephenson JR (1983) Homology between phosphotyrosine acceptor site of human c-*abl* and viral oncogene products. Nature 304: 167–169

Groffen J, Stephenson JR, Heisterkamp N, deKlein A, Bartram C, Grosveld G (1984) Philadelphia chromosomal breakpoints are clustered within a limited region, *bcr*, on chromosome 22. Cell 36: 93–99

Grosveld G, Verwoerd T, Van Agthoven T, deKlein A, Ramachandran KL, Heisterkamp N, Stam K, Groffen J (1986) The chronic myelocytic cell line K562 contains a breakpoint in *bcr* and produces a chimeric *bcr-abl* transcript. Mol Cell Biol 6: 607–616

Harris AW, Langdon WY, Alexander WS, Hariharan IK, Rosenbaum H, Vaux D, Webb E, Bernard O, Crawford M, Abud H, Adams JM, Cory S (1988) Transgenic mouse models for hematopoietic tumorigenesis. Curr Top Microbiol Immunol 141: 82–93

Heisterkamp N, Stephenson JR, Groffen J, Hansen PF, deKlein A, Bartram C, Grosveld G (1983) Localization of the c-*abl* oncogene adjacent to a translocation breakpoint in chronic myelocytic leukemia. Nature 306: 239–242

Heisterkamp N, Stam K, Groffen J, deKlein A, Grosveld G (1985) Structural organization of the *bcr* gene and its role in the Ph′ translocation. Nature 315: 758–761

Henderson LE, Krutzsch HC, Oroszlan S (1983) Myristyl amino-terminal acylation of murine retrovirus proteins: an unusual post-translational protein modification. Proc Natl Acad Sci USA 80: 339–343

Henkemeyer MJ, Gertler FB, Goodman W, Hoffmann FM (1987) The *Drosophila* Abelson proto-oncogene homolog: identification of mutant alleles that have pleiotropic effects late in development. Cell 51: 821–828

Henkemeyer MJ, Bennett RL, Gertler FB, Hoffmann FM (1988) DNA sequence, structure, and tyrosine kinase activity of the *Drosophila melanogaster* Abelson proto-oncogene homolog. Mol Cell Biol 8: 843–853

Hermans A, Heisterkamp N, Von Lindern M, Van Baal S, Meijer D, Van der Plas D, Wiedemann LM, Groffen J, Bootsma D, Grosveld G (1987) Unique fusion of *bcr* and c-*abl* genes in Philadelphia positive acute lymphoblastic leukemia. Cell 51: 33–40

Hoffmann FM, Fresco LD, Hoffmann-Falk H, Shilo B-Z (1983) Nucleotide sequences of the *Drosophila src* and *abl* homologs: conservation and variability in the *src* family oncogenes. Cell 35: 393–401

Kamps MP, Taylor SS, Sefton BM (1984) Direct evidence that oncogenic tyrosine kinases and cyclic AMP-dependent protein kinase have homologous ATP-binding sites. Nature 310: 589–592

Kipreos ET, Wang JY-J (1988) Reversible dependence on growth factor interleukin-3 in myeloid cells expressing temperature sensitive v-*abl* oncogene. Oncogene Res 2 277–284

Kipreos ET, Lee GL, Wang JY-J (1987) Isolation of temperature-sensitive tyrosine kinase mutants of the v-*abl* oncogene by screening with antibodies for phospho-tyrosine. Proc Natl Acad Sci USA 84: 1345–1349

Kloetzer WR, Kurzrock L, Smith M, Talpaz M, Spiller J, Gutterman M, Arlinghaus R (1985) The human cellular *abl* gene product in the chronic myelogenous leukemia cell line has an associated tyrosine protein kinase activity. Virology 140: 230–238

Konopka JB, Witte ON (1985) Detection of c-*abl* tyrosine kinase activity *in vitro* permits direct comparison of normal and altered *abl* gene products. Mol Cell Biol 5: 3116–3123

Konopka J, Watanabe S, Witte ON (1984) An alteration of the human c-*abl* protein in K562 leukemia cells unmasks associated tyrosine kinase activity. Cell 37: 1035–1042

Kurzrock R, Shtalrid M, Romero P, Kloetzer WS, Talpas M, Trujillo JM, Blick M, Beran M, Gutterman JU (1987) A novel c-*abl* protein product in Philadelphia-positive acute lymphoblastic leukaemia. Nature 325: 631–635

Latt SA, Goff SP, Tabin CJ, Paskind M, Wang JY-J, Baltimore D (1983) Cloning and analysis of reverse transcript P160 genomes of Abelson murine leukemia virus. J Virol 45: 1195–1199

Lee R, Paskind M, Wang JY-J, Baltimore D (1985) In: Weiss R, Teich B, Varmus H, Coffin J (eds) RNA tumor viruses. Cold Spring Harbor Laboratory, Cold Spring Harbor, NY, pp 861–868

Mathey-Prevot B, Baltimore D (1985) Specific transforming potential of oncogenes encoding protein kinases. EMBO J 4: 1769–1774

Mathey-Prevot B, Baltimore D (1988) Recombinants within the tyrosine kinase region of v-*abl* and v-*src* identify a v-*abl* segment that confers lymphoid specificity. Mol Cell Biol 8: 234–240

Mathey-Prevot B, Nabel G, Palacios R, Baltimore D (1986) Abelson virus abrogation of interleukin-3 dependence in a lymphoid cell line. Mol Cell Biol 6: 4133–4135

Mayer BJ, Hamaguchi M, Hanafusa H (1988) A novel viral oncogene with structural similarity to phospholipase C. Nature 332: 272–275

McCubrey J, Holland GD, McKearn J, Risser R (1989) Abrogation of factor dependence in 2 IL-3-dependent lines can occur by autocrine and nonautocrine mechanisms. Oncogene Res 4: 97–109

McLaughlin J, Chianese E, Witte ON (1987) *In vitro* transformation of immature hematopoietic cells by the P210 BCR/ABL oncogene product of the Philadelphia chromosome. Proc Natl Acad Sci USA 84: 6558–6562

Maijer D, Hermans A, Von Lindern M, Van Agthoven T, deKlein A, Mackenbush P, Grootegoed A, Talarico D, Della Valle G, Grosveld G (1987) Molecular characterization of the testis specific c-*abl* mRNA in mouse. EMBO J 6: 4041–4048

Mes-Masson A-M, McLaughlin J, Daley GQ, Paskind M, Witte ON (1986) Overlapping cDNA clones define the complete coding region for the $P210^{c\text{-}abl}$ gene product associated with chronic myelogenous leukemia cells containing the Philadelphia chromosome. Proc Natl Acad Sci USA 83: 9768–9772

Muller R, Slamon DJ, Tremblay JM, Cline MJ, Verma IM (1982) Differential expression of cellular oncogenes during pre- and postnatal development of the mouse. Nature 299: 640–644

Murtagh K, Skladany G, Hoag J, Rosenberg N (1987) Abelson murine leukemia virus variants with increased oncogenic potential. J Virol 60: 599–606

Nowell PC, Hungerford DA (1960) A minute chromosome in human chronic granulocytic leukemia. Science 132: 1497–1499

Ohno S, Migita S, Wiener F, Babonits M, Klein G, Mushinski JF, Potter M (1984) Chromosomal translocations activating *myc* sequences and transduction of v-*abl* are critical events in the rapid induction of plasmacytomas by pristane and Abelson virus. J Exp Med 159: 1762–1777

Oliff A, Agranovsky O, McKinney MD, Murty VVVS, Buchwitz R (1985) Friend murine leukemia virus-immortalized myeloid cells are converted into tumorigenic cell lines by Abelson leukemia virus. Proc Natl Acad Sci USA 82: 3306–3310

Oppi C, Shore SK, Reddy EP (1987) Nucleotide sequence of testis-derived c-*abl* cDNAs: implications for testis-specific transcription and *abl* oncogene activation. Proc Natl Acad Sci USA 84: 8200–8204

Palumbo GJ, Ozanne B, Kettman JR (1989) Growth factors increase the efficiency of transformation of bone marrow cells by Abelson murine leukemia virus. (In press)

Pendergast AM, Traugh JA, White ON (1987) Normal cellular and transformation-associated *abl* proteins share common sites for protein kinase C phosphorylation. Mol Cell Biol 7: 4280–4289

Pierce JH, DiFiore PP, Aaronson SA, Potter M, Pumphrey J, Scott A, Ihle JN (1985) Neoplastic transformation of mast cells by Abelson-MuLV: abrogation of IL-3 dependence by a nonautocrine mechanism. Cell 41: 685–693

Poirier Y, Kozak C, Jolicoeur P (1988) Identification of a common helper provirus integration site in Abelson virus-induced lymphoma DNA. J Virol 62: 3985–3992

Ponticelli AS, Whitlock CA, Rosenberg N, Witte ON (1982) *In vivo* tyrosine phosphorylations of the Abelson virus transforming protein are absent in its normal cellular homolog. Cell 29: 953–960

Potter M, Sklar MD, Rowe WP (1973) Rapid viral induction of plasmacytomas in pristane-primed BALB/c mice. Science 182: 592–594

Premkumar E, Potter M, Singer PA, Sklar M (1975) Synthesis, surface disposition, and secretion of immunoglobulins by Abelson virus-transformed lymphosarcoma cell lines. Cell 6: 149–159

Prywes R, Foulkes JG, Rosenberg N, Baltimore D (1983) Sequences of the A-MuLV protein needed for fibroblast and lymphoid cell transformation. Cell 34: 569–579

Prywes R, Foulkes JG, Baltimore D (1985a) The minimum transforming region of v-*abl* is the segment encoding protein-tyrosine kinase. J Virol 54: 114–122

Prywes R, Hoag J, Rosenberg N, and Baltimore D (1985b) Protein stabilization explains the *gag* requirement for transformation of lymphoid cells by Abelson murine leukemia virus. J Virol 54: 124–132

Raschke WC, Baird S, Ralph P, Nakoinz I (1978) Functional macrophage cell lines transformed by Abelson murine leukemia virus. Cell 15: 261–267

Reddy EP, Smith MJ, Srinivasan A (1983) Nucleotide sequence of Abelson murine leukemia virus genome: structural similarity of its transforming gene product to other *onc* gene products with tyrosine specific kinase. Proc Natl Acad Sci USA 80: 3623–3627

Rees-Jones RW, Goldfarb M, Goff SP (1989) Abelson murine leukemia virus induces PDGF-independent fibroblast growth: correlation with kinase activity and dissociation from full morphologic transformation. Mol Cell Biol 9: 278–287

Renshaw MW, Capozza MA, Wang JY-J (1988) Differential expression of type specific c-*abl* mRNAs in mouse tissues and cell lines. Mol Cell Biol 8: 4547–4551

Reynolds EH, Sacks TL, Deobagkar DN, Stephenson JR (1978) Cells nonproductively transformed by Abelson murine leukemia virus express a high molecular weight polyprotein containing structural and nonstructural components. Proc Natl Acad Sci USA 75: 3974–3978

Risser R (1982) The pathogenesis of Abelson virus lymphomas of the mouse. Bioch Biophys Acta 651: 213–244

Risser R, Pollack R (1974) A non-selective analysis of SV40 transformation of mouse 3T3 cells. Virology 59: 477–489

Risser R, Potter M, Rowe WP (1978) Abelson virus induced lymphomagenesis in mice. J Exp Med 148: 714–726

Risser R, Kaehler D, Lamph WW (1985) Different genes control the susceptibility of mice to Moloney or Abelson murine leukemia viruses. J Virol 55: 547–553

Rosenberg N, Baltimore D (1976) A quantitative assay for transformation of bone marrow cells by Abelson murine leukemia virus. J Exp Med 143: 1453–1463

Rosenberg N Baltimore D (1978) The effect of helper virus on Abelson virus-induced transformation of lymphoid cells. J Exp Med 147: 1126–1141

Rosenberg NE, Witte ON (1980) Abelson murine leukemia virus mutants with alterations in the virus-specific P120 molecule. J Virol 33: 340–348
Rosenberg N, Baltimore D, Scher CD (1975) In vitro transformation of lymphoid cells by Abelson murine leukemia virus. Proc Natl Acad Sci USA 72: 1932–1936
Rovera G, Valtieri M, Mavillo F, Reddy EP (1987) Effect of Abelson murine leukemia virus on granulocytic differentiation and interleuken-3 dependence of a murine progenitor cell line. Oncogene 1: 29–35
Rowley JD (1973) A new consistent chromosomal abnormality in chronic myelogenous leukemia identified by quinacrine fluorescence and giemsa staining. Nature 243: 290–291
Rubin CM, Carrino JJ, Dickler MN, Leibowitz D, Smith SD, Westbrook CA (1988) Heterogeneity of genomic fusion of BCR and ABL in Philadelphia chromosome-positive acute lymphoblastic leukemia. Proc Natl Acad Sci USA 85: 2795–2799
Sadowski I, Stone JC, Pawson T (1986) A noncatalytic domain conserved among cytoplasmic protein-tyrosine kinases modifies the kinase function and transforming activity of Fujinami sarcoma virus $P130^{gag, fps}$. Mol Cell Biol 6: 4396–4408
Savard P, DesGroseillers L, Rassart E, Poirier Y, Jolicoeur P (1987) Important role of the long terminal repeat of the helper Moloney murine leukemia virus in Abelson virus-induced lymphoma. J Virol 61: 3266–3275
Scher CD (1978) Effect of pseudotype on Abelson virus and Kirsten sarcoma virus-induced leukemia. J Exp Med 147: 1044–1053
Scher CD, Siegler R (1975) Direct transformation of 3T3 cells by Abelson murine leukemia virus. Nature 253: 729–731
Scott ML, Davis MM, Feinberg MB (1986) Transformation of T-lymphoid cells by Abelson murine leukemia virus. J Virol 59: 434–443
Sefton B, Hunter T, Raschke WC (1981) Evidence that the Abelson virus protein functions *in vivo* as a protein kinase that phosphorylates tyrosine. Proc Natl Acad Sci USA 78: 1552–1556
Serunian LA, Rosenberg N (1986) Abelson virus potentiates long-term growth of mature B lymphocytes. Mol Cell Biol 6: 183–194
Shinnick TM, Lerner RA, Sutcliffe JG (1981) Nucleotide sequence of Moloney murine leukaemia virus. Nature 293: 543–548
Shtivelman E, Lifshitz B, Gale RP, Canaanii E (1985) Fused transcript of *abl* and *bcr* genes in chronic myelogenous leukaemia. Nature 315: 550–552
Shtivelman E, Lifshitz B, Gale RP, Roe BA, and Canaani, E (1986) Alternative splicing of RNAs transcribed from the human *abl* gene and from the *bcr-abl* fused gene. Cell 47: 277–284
Siden EJ, Baltimore D, Clark D, Rosenberg NE (1979) Immunoglobulin synthesis by lymphoid cells transformed *in vitro* by Abelson murine leukemia virus. Cell 16: 389–396
Siden E, Alt FW, Shinefeld L, Sato V, Baltimore D (1981) Synthesis of immunoglobulin mu chain gene products precedes synthesis of light chains during B-lymphocyte development. Proc Natl Acad Sci USA 78: 1823–1827
Siegler R, Zajdel S, Lane I (1972) Pathogenesis of Abelson-virus-induced murine leukemia. J Natl Cancer Inst 48: 189–218
Sklar MD, Shevach EM, Green I, Potter M (1975) Transplantation and preliminary characterization of lymphocyte surface markers of Abelson virus induced lymphoma. Nature 253: 550–552
Spolski R, Miescher G, Erard F, Risser R, MacDonald HR, Mak TW (1988) Regulation of expression of T cell gamma chain, L3T4, and Lyt 2 messages in Abelson/Moloney virus-transformed T cell lines and thymocytes. Eur J Immunol 18: 295–300
Srinivasan A, Dunn CY, Yuasa Y, Devare SG, Reddy EP, Aaronson SA (1982) Abelson murine leukemia virus: structural requirements for transforming gene function. Proc Natl Acad Sci USA 79: 5508–5512
Stahl ML, Ferenz CR, Kelleher KL, Kriz RW, Knopf JL (1988) Sequence similarity of phospholipase C with the non-catalytic region of *src*. Nature 332: 269–272
Takemori T, Miyazoe I, Shirasawa T, Taniguchi M, Graf T (1987) A temperature-sensitive mutant of Abelson murine leukemia virus confers inducibility of IgM expression to transformed lymphoid cells. EMBO J 6: 951–956

Tidmarsh GF, Heimfeld S, Spangrude GJ, Weissman IL, Muller-Sieberg CE (1989) Identification of a novel bone marrow derived B cell progenitor population which coexpresses B220 and Thy-1 and is highly enriched for Abelson leukemia virus targets. Mol Cell Biol (in press)

Vogt M, Lesley J, Bogenberger JM, Haggblom C, Swift S, Haas M (1988) The induction of growth factor-independence in murine myelocytes by oncogenes results in monoclonal cell lines and is correlated with cell crisis and karyotic instability. Oncogene Res 2: 49–63

Walker LC, Ganesan TS, Dhut S, Gibbons B, Lister TA, Rothbard J, Young BD (1987) Novel chimeric protein expressed in Philadelphia positive acute lymphoblastic leukaemia. Nature 329: 851–853

Wang JY-J (1988) Negative regulation of c-*abl* tyrosine kinase by its variable N-terminal amino acids. Oncogene Res 3: 293–298

Wang JY-J, Queen C, Baltimore D (1982) Expression of an Abelson murine leukemia virus encoded protein in *Escherichia coli* caused extensive phosphorylation of tyrosine residues. J Biol Chem 257: 13181–13184

Wang JY-J, Ledley F, Goff S, Lee R, Groner Y, Baltimore D (1984) The mouse c-*abl* locus: molecular cloning and characterization. Cell 36: 349–356

Watanabe S, Witte ON (1983) Site directed deletions of Abelson murine leukemia virus define 3′ sequences essential for transformation and lethality. J Virol 45: 1028–1036

Westin EH, Wong-Staal F, Gelman EP, Dalla-Favera R, Papas T, Lautenberger JA, Eva A, Reddy EP, Tronick SR, Aaronson SA, Gallo RC (1982) Expression of cellular homologues of retroviral oncogenes in human hematopoietic cells. Proc Natl Acad Sci USA 79: 2490–2494

Whitlock CA, Witte ON (1981) Abelson virus-infected cells can exhibit restricted *in vitro* growth and low oncogenic potential. J Virol 40: 577–584

Whitlock CA, Ziegler S, Witte ON (1983) Progression of the transformed phenotype in clonal lines of Abelson virus-infected lymphocytes. Mol Cell Biol 3: 596–604

Witte ON, Rosenberg N, Paskin M, Shields A, Baltimore D (1978) Identification of an Abelson murine leukemia virus encoded protein present in transformed fibroblasts and lymphoid cells. Proc Natl Acad Sci USA 75: 2488–2492

Witte ON, Goff S, Rosenberg N, Baltimore D (1980) A transformation-defective mutant of Abelson murine leukemia virus lacks protein kinase activity. Proc Natl Acad Sci USA 77: 4993–4997

Eukaryotic RAS Proteins and Yeast Proteins with Which They Interact

Daniel Broek

1 Background

1.1 Nutrients and Growth Control in Yeast

Wild type *Saccharomyces cerevisiae* cells coordinate cell division with the availability of essential nutrients. In rich media, cells divide continuously. If the medium is moderately limited for some nutrients required for growth and division, cells grow slowly, requiring more time for each cell cycle. The increase in cell cycle time is due to a preferential increase in the time required to traverse the G1 phase of the cell cycle (Johnston et al. 1977). If the medium is severely limited for nutrients, haploid cells arrest in G1 in an altered physiologic state. In this altered state, the cells are resistant to heat shock and can survive long periods of nutrient deprivation (Pringle and Hartwell 1981). Diploid cells severely limited for nutrients can arrest in G1 and sporulate (Esposito and Klapholz 1981). These data suggest that nutritional information may influence the nature of events in G1.

Kenneth Norris Cancer Hospital and Research Institute, University of Southern California, School of Medicine, Los Angeles, CA 90033, USA

Current Topics in Microbiology and Immunology, Vol. 147

In 1974, L. Hartwell introduced the use of mutants to examine cell cycle stage-specific events in yeast. More than 60 such mutants have been isolated (Pringle and Hartwell 1981). These mutants, termed cell division cycle or *cdc* mutants, are defined by mutations that cause a yeast strain to arrest at a specific stage of the cell cycle (Hartwell 1974). A number of temperature-sensitive (ts) conditional lethal mutations exist that arrest strains in G1. These mutations have been placed into two groups on the basis of morphological cell cycle markers (called landmark events) and mutant phenotypes in the arrested state (Pringle and Hartwell 1981). Cells treated with mating pheromone arrest in G1 and continue to grow, increasing in size and forming a characteristically distorted cell shape, the "schmoo". At the restrictive temperature (usually 37 °C) ts *cdc28*[1], *cdc36*, *cdc37*, and *cdc39* mutants arrest in G1 as schmoos, and are competent for mating. In contrast, the ts *cdc19*, *cdc25*, *cdc33*, and *cdc35* mutants arrest in G1, do not continue to grow, and cannot mate (Pringle and Hartwell 1981). By examining landmark events in strains arrested with ts *cdc19*, *cdc25*, *cdc33*, and *cdc35* alleles, it is likely that the mutants in this class arrest earlier in G1 than those in the *cdc28* class (Pringle and Hartwell 1981). Significantly, mutants in the *cdc19* class arrest in a state resembling that of nutrient-limited cells (Pringle and Hartwell 1981). Together with the evidence described below, this observation makes the genes of the *cdc19* class candidates for genes whose products signal the nutritional state of the cell or its environment.

1.2 cAMP and Growth Control in Yeast

Matsumoto and colleagues isolated cAMP-permeable strains of yeast that require cAMP for growth (Matsumoto et al. 1982). When deprived of exogenously added cAMP, these mutants arrest in G1 in a state very similar to that of mutants in the *cdc19* class or of nutritionally starved cells (Matsumoto et al. 1983). The cAMP-requiring mutants isolated by Matsumoto et al. contained mutations in three genes, *CYR1*, *CYR2*, and *CYR3*. *CDC35* is allelic with *CYR1* (Boutlet et al. 1985). Biochemical analysis indicated that *cyr1* mutants, but not *cyr2* or *CYR3* mutants, lack adenylate cyclase activity (Matsumoto et al. 1982). This and other evidence led Matsumoto et al. (1984) to suggest that *CYR1* was the structural gene for adenylate cyclase. This suggestion was later confirmed when the *CYR1* locus was cloned and characterized (Casperson et al. 1985; Boutelet et al. 1985; Kataoka et al. 1985b). Biochemical evidence indicated that the *cyr2* and *CYR3* mutations affect the cAMP-dependent protein kinase (cAMPdPK) (Uno et al. 1984b). The *cyr2* mutation is allelic with *CDC25* (Matsumoto, unpublished data).

Matsumoto et al. (1982) also identified a suppressor mutation of *cyr1*, called *bcy1* (bypass cylic AMP). Cells with the *bcy1* mutation are viable in the absence of detectable adenylate cyclase activity. The cAMPdPK activity in *bcy1* cells is largely cAMP independent (Matsumoto et al. 1982). Using a photoactivatable analog of

[1] *S. cerevisiae* genetics nomenclature refers to wild-type genes and dominant mutants using capital letters; recessive mutations are referred to using lower case letters. The mammalian *ras* gene is referred to using lower case letters.

cAMP, 8-azido-cAMP, UNO et al. (1982) showed that *bcy1* cells have reduced levels of a cAMP-binding protein, which they presumed was the regulatory subunit of the cAMPdPK (UNO et al. 1982; MATSUMOTO et al. 1982). *BYC1* is the structural gene for the regulatory subunit of the cAMPdPK. Evidence for this comes from two sources. First, the cloning and sequencing of *BCY1* revealed it to share significant amino acid sequence homology with the regulatory subunit of the vertebrate cAMP-dPK (CANNON and TATCHELL 1987; TODA et al. 1987b). Second, the *BCY1* protein purified from an *E. coli* expression system can inhibit the activity of the catalytic subunit of the yeast cAMPdPK and this inhibition was abolished in the presence of exogenously added cAMP (HIXSON and KREBS 1980; JOHNSON et al. 1987; KURET et al. 1988).

The phenotypes apparent in cells containing mutant components of the cAMP pathway suggest that cAMP may have some role in the response to nutrient limitation (see SHIN et al. 1987). In general, yeast strains with mutations that reduce the activity of the cAMP pathway respond as if nutrient limited, even under rich growth conditions. For example, both the *cyr1* mutation and limitation for essential nutrients lead to similar arrests in G1 (MATSUMOTO et al. 1983). Sporulation is also defective in strains with reduced activity of the cAMP pathway. Wild-type diploid cells adapted for growth on a nonfermentable carbon source sporulate when starved for a nitrogen source (ESPOSITO and KLAPHOLZ 1981). Diploid cells with debilitated components of the cAMP pathway, for example cells homozygous for ts mutations of *CYR1*, sporulate prematurely. Such cells will form spores on nitrogen-rich, acetate-based media (SHILO et al. 1978; MATSUMOTO et al. 1983). Yeast strains with reduced activity of the cAMP pathway also accumulate abnormally large amounts of the storage carbohydrates glycogen and trehalose (TATCHELL et al. 1985; TODA et al. 1985). These defects contrast with those of strains having enhanced activity of the cAMP pathway. Such strains fail to respond to nutrient limitation. For example, cells containing the *bcy1* mutation, which results in a constitutively active cAMPdPK, fail to arrest in G1 when deprived of nutrients (MATSUMOTO et al. 1983), and do not sporulate (MATSUMOTO et al. 1983) or accumulate storage carbohydrates (UNO et al. 1984a; CANNON et al. 1986) under any conditions. On the basis of these data, Matsumoto and coworkers proposed that cAMPdPK activity promotes mitotic growth, and loss of cAMPdPK activity promotes the alternate pathways of G1 arrest, and meiosis and sporulation (MATSUMOTO et al. 1983). Work with yeast homolog of the vertebrate *ras* oncogene implicated the yeast *RAS* genes as controlling elements of adenylate cyclase in yeast, and prompted a more detailed analysis of the cAMP pathway.

2 Yeast *RAS* Function

2.1 The Yeast *RAS* Proteins Are Structurally Homologous to the Human *ras* Proteins

Despite extensive effort by many investigators, the effector system modulated by the vertebrate *ras* proteins to control cell growth remains unknown (BARBACID, 1987).

Therefore, in an effort to understand the mechanism of *ras* proteins in vertebrate cells, and the role of mutant forms of these proteins in tumorigenesis, homolog of the vertebrate *ras* proteins were identified in the yeast *Saccharomyces cerevisiae*. *Saccharomyces cerevisiae* has two genes, *RAS1* and *RAS2*, homologous to the mammalian *ras* genes (DeFeo-Jones et al. 1983; Powers et al. 1984). Both genes were identified by screening a yeast DNA library at low hybridization stringency with a vertebrate *ras* complementary (c) DNA probe. As would be expected of genes with similar DNA sequences, the proteins encoded by the yeast *RAS* genes are similar to the vertebrate *ras* proteins. The mammalian H-*ras* gene encodes a protein of 189 amino acids (Taparowsky et al. 1982). Nearly 90% of the first 80 amino acids of the H-*ras* protein are conserved in both yeast *RAS* proteins, and nearly 50% of the next 80 amino acids are conserved (Powers et al. 1984). The two yeast proteins are very similar to one another. Approximately 90% of the first 180 amino acids are identical in *RAS1* and *RAS2*. The yeast *RAS* proteins are larger than the mammalian *ras* proteins. *RAS1* and *RAS2* are 309 and 322 amino acids respectively (DeFeo-Jones et al. 1983; Powers et al. 1984). The greater size of the yeast proteins is due largely to an insertion at position 180 of about 120 amino acids. There are also an additional seven amino acids at the amino terminus of the yeast proteins (DeFeo-Jones et al. 1983; Powers et al. 1984). All *ras* proteins, including those from yeast, terminate with the sequence Cys-A-A-X, where A is an aliphatic amino acid and X is not restricted (Shimizu et al. 1983; Taparowsky et al. 1983; Powers et al. 1984). The carboxy-terminal cysteine in this sequence is required for proper membrane localization and function of the *RAS* proteins, perhaps by serving as the site for attachment of a fatty acid (Powers et al. 1986; Tamanori et al. 1986; Deschenes and Broach 1987).

The human *ras* proteins and the yeast *RAS* proteins have similar intrinsic biochemical properties. The mammalian *ras* proteins are guanine nucleotide binding proteins which are capable of hydrolyzing GTP (Papageorge et al. 1982; Gibbs et al. 1984; Sweet et al. 1984; McGrath et al. 1984). The yeast *RAS* proteins are similar to their mammalian homologs in their ability to bind and hydrolyze GTP (Tamanori et al. 1985). Further, the mutant *RAS2*$^{val\,19}$ protein, like it mammalian counterpart H-*ras*$^{val\,12}$, has a markedly reduced GTPase activity (Tamanori et al. 1985).

2.2 The Yeast *RAS* Proteins Are Functionally Homologous to the Human *ras* Proteins

In addition to being structurally similar to the mammalian *ras* genes, the yeast *RAS* genes are functionally similar. In yeast, chromosomal genes can be accurately replaced by recombinant gene constructs. Using this technique, it was shown that while neither *RAS1* nor *RAS2* was itself essential, some *RAS* function is required for spore germination and cell viability (Kataoka et al. 1984; Tatchell et al. 1984; Temeles et al. 1984). This essential function can be supplied by overexpression of the H-*ras* gene in a cell lacking endogenous yeast *RAS* genes (Kataoka et al. 1985a). Conversely, the yeast *RAS* genes can also function in mammalian cells. DeFeo-Jones and coworkers demonstrated that a modified yeast *RAS1* gene can induce morphological transformation of NIH 3T3 cells (DeFeo-Jones et al. 1985).

The finding that the human H-*ras* protein could serve *RAS* function in yeast and the yeast *RAS1* protein could function in mammalian cells had dramatic implications. These observations demonstrated a conservation of protein-protein interactions (involving *RAS*) from yeast to man. Further, these studies imply the study of the yeast *RAS* protein-protein interactions should give insight into mammalian *ras* protein-protein interactions. Consequently, the study of yeast *RAS* turned to identifying the proteins in yeast which interact directly with *RAS* (both yeast and human) proteins.

2.3 *RAS* Proteins Regulate cAMP Levels in Yeast

The study of the yeast *RAS* genes was greatly aided by studying the phenotypes of cells harboring mutant *RAS* genes. These phenotypes turned out to be remarkably similar to the phenotypes observed for mutants in the cAMP pathway as well as *cdc* mutants causing growth arrest.

Cell growth is profoundly disrupted in mammalian cells containing mammalian *ras* genes with a mutation of the glycine at position 12 (H-$ras^{val\,12}$). An analogous mutation was made in the yeast *RAS2* gene ($RAS2^{val\,19}$). In a striking indication of the extent of functional conservation between yeast and mammalian *ras* genes, when the "activated" $RAS2^{val\,19}$ was introduced into yeast, it acted dominantly to alter the growth properties of the cells (Kataoka et al. 1984). Initially, it was found that $RAS^{val\,19}$ mutant cells do not sporulate, and are poorly viable when stored (Kataoka et al. 1984). These phenotypes are similar to those described for cells containing the *bcy1* mutation (Matsumoto et al. 1983), and suggested a link between *RAS* and cAMP in yeast. Yeast strains with mutant *RAS* genes were examined to determine whether they shared any other phenotypes with strains containing mutant components of the cAMP pathway. The results obtained supported a link between the *RAS* genes and cAMP. Diploid yeast strains with reduced levels of adenylate cyclase activity, unlike wild-type yeast, sporulate on nitrogen-rich medium containing acetate (YPA). Diploid cells lacking a *RAS2* gene (but containing *RAS1*) also sporulate on YPA (Toda et al. 1985). Wild-type yeast cells limited for nutrients accumulate the storage carbohydrates glycogen and trehalose (Lillie and Pringle 1980), but strains containing the $RAS2^{val\,19}$ mutation do not (Toda et al. 1985). The enzyme trhalase, which degrades trehalose, is a substrate of the cAMPdPK in yeast and is activated by phosphorylation (Uno et al. 1984a). $RAS2^{val\,19}$ strains have elevated trehalase activity that cannot be significantly increased by cAMPdPK phosphorylation in vitro (Toda et al. 1985). All these data taken together prompted investigators to test whether *RAS* is a regulator of cAMP levels.

Disruption of both *RAS* genes (*RAS1* and *RAS2*) in a strain is lethal (Kataoka et al. 1984; Tatchell et al. 1984). As described above, the *bcy1* mutation was originally isolated as a suppressor of cells lacking adenylate cyclase activity (Matsumoto et al. 1982). Toda and coworkers showed the *bcy1* mutation can also suppress loss of *RAS* function (Toda et al. 1985). This observation strongly supported the idea that the *RAS* proteins modulate cAMP levels in yeast and prompted the biochemical investigation to determine the role *RAS* proteins play in cAMP metabolism.

Examination of the effect of mutant *RAS* genes on intracellular cAMP levels suggested they control cAMP levels in yeast. Relative to wild-type strains, the levels of cAMP are increased approximately fourfold in strains containing the *RAS2*$^{val\,19}$ mutation, and decreased approximately fourfold in strains lacking the *RAS2* gene (Toda et al. 1985). Strains lacking both *RAS* genes, but which were viable because they contained the *bcy1* mutation, did not contain detectable cAMP (Toda et al. 1985).

2.4 *RAS* Proteins Regulate the Yeast Adenylate Cyclase

Casperson et al. had shown that the in vitro adenylate cyclase activity of *S. cerevisiae* membranes could be stimulated by addition of a nonhydrolyzable GTP

Table 1. Adenylate cyclase activity in membranes

Genotype[b]	Protein added[c]	Assay conditions[a]		
		Mn^{2+}	Mg^{2+}	Mg^{2+}, Gpp(NH)p
1. *RAS1 RAS2*	—	49.2	3.7	14.3
2. *RAS1 RAS2*^{val19}	—	51.9	14.2	16.1
3. *ras1*$^-$ *ras2*$^-$ *bcy1*	—	51.0	1.1	0.7
4. *RAS1 RAS2 cyr1-1*	—	0.3	0.1	0.2
5. *ras1*$^-$ *ras2*$^-$ *bcy1* + *RAS1 RAS2 cyr-1*	—	29.5[c]	2.3[c]	23.5[c]
6. *ras1*$^-$ *ras2*$^-$ *bcy1*	*RAS2*[e]	57	3.1	72
7. *ras1*$^-$ *ras2*$^-$ *bcy1*	f-*RAS1*[e]	57	0.4	21.3
8. *ras1*$^-$ *ras2*$^-$ *bcy1*	H-*ras*[e]	57	0.7	19.3
9. *cdc25*$^-$, *pCDC25*	—	56	4.1	10.4
10. *cdc25*$^-$, *pTPK1*[f]	—	39	0.2	9.4
11. *cdc25*$^-$, *pRAS2*^{val19}	—	71	4.0	8.4

[a] Membranes from the indicated strains were prepared and adenylate cyclase was assayed as described (Toda et al. 1985). Membranes were assayed either in the presence of 2.5 m*M* Mn^{2+} or 2.5 m*M* Mg^{2+} or 2.5 m*M* Mg^{2+} and 10 μ*M* Gpp(NH)p. Adenylate cyclase activity is expressed in units of picomoles of cAMP generated per milligram of membrane protein per minute. Mn^{2+}-dependent activity reflects the amount of catalytic adenylate cyclase in a membrane preparation and this activity is not highly dependent on auxillary proteins.

[b] The full genotypes of the indicated strains are given in Toda et al. (1985) and Broek et al. (1987).

[c] See Broek et al. (1985) for description of preparation of these proteins. *RAS2* indicates the *RAS2* protein from *S. cerevisiae*. f-*RAS1* is a fusion protein of the *RAS1* protein from *S. cerevisiae*. H-*ras* indicates the human H-*ras* protein.

[d] In this experiment membranes from the indicated strains were incubated wither alone or together for 2 h at 0 °C in 25 m*M* MES (pH 6.2), 1 m*M* ATP, 0.06% Lubrol, with or without 30 μ*M* Gpp(NH)p. They were diluted as described such that the final Lubrol concentration was 0.01%. Membranes were then incubated at 15 °C for 60 min and assayed as before (Broek et al. 1985)

[e] The *RAS* proteins purified from *E. coli* expression system are bound to GDP. For the experiments under Mg^{2+}, Gpp(NH)p, the Gpp(NH)p was preincubated with the indicated *ras* protein prior to mixing with the membranes.

[f] *TPK1* is a gene for a yeast cAMPdPK (Toda et al. 1987c).

analog, Gpp(NH)p (see *RAS1 RAS2* in Table 1, line 1) (CASPERSON et al. 1983). Toda and coworkers tested whether the *RAS* protein is the GTP-binding protein responsible for the GTP-dependent stimulation of adenylate cyclase. Adenylate cyclase activity in isolated membranes was measured from various *RAS* mutant strains. Strains lacking *RAS* genes, but viable due to a *bcy1* mutation, had virtually undetectable levels of adenylate cyclase activity in the presence of Mg^{2+} and Gpp(NH)p (see Table 1, line 3). In contrast, adenylate cyclase activity from $RAS2^{val\,19}$ strains was elevated and could not be stimulated by the addition of Gpp(NH)p (Table 1, line 2) (TODA et al. 1985). Finally, in a striking demonstration of the link between *RAS* and adenylate cyclase, a GTP-responsive adenylate cyclase was reconstituted by mixing membranes from cell lacking a functional adenylate cyclase (*RAS1 RAS2 cyr1*) and membranes from cell lacking *RAS* proteins ($ras1^-ras2^-bcy1$ cells) (Table 1, lines 3–5); (TODA et al. 1985). Each of these membranes alone lacks Mg^{2+}-GTP-dependent adenylate cyclase activity but when combined is fully GTP responsive. These results conclusively established that *RAS* proteins are controlling elements of adenylate cyclase in yeast.

Scolnick and coworkers showed that Harvey *ras* proteins possess an intrinsic guanine nucleotide binding activity (PAPAGEORGE et al. 1982). This finding suggested the *ras* proteins might function in a manner analogous to $G_s\alpha$ of the mammalian adenylate cyclase system which activates adenylate cyclase when bound to GTP but not when bound to GDP (see GILMAN 1987). The GTPase activity of *ras* proteins thus would serve as an internal "turn-off" mechanism. This model is particularly attractive because most oncogenic *ras* proteins have a reduced GTPase activity. Consequently, the model predicts the oncogenic *ras* proteins would be bound to GTP at inappropriate times and thus constitutively activate their target enzyme. The study of the human and yeast *ras* proteins' ability to activate the yeast adenylate cyclase gene gave the first biochemical data favoring this model of *ras* action. The membrane-bound yeast adenylate cyclase activity can be stimulated in vitro by yeast and human *ras* proteins bound to GTP but not those bound to GDP (see Table 1, lines 3, 6–8; BROEK et al. 1985; FIELD et al. 1987).

Several lines of evidence indicate that *RAS* proteins interact directly with the yeast adenylate cyclase molecule. First, two groups have expressed portions of the yeast adenylate cyclase in *E. coli* and found that addition of *RAS* proteins bound to GTP activate the adenylate cyclase (UNO et al. 1985; T. KATOAKA, personal communication). Second, the yeast adenylate cyclase has been purified by an epitope addition immunoaffinity purification procedure. The highly purified adenylate cyclase can be dramatically stimulated by the human and yeast *RAS* proteins bound to GTP (FIELD et al. 1987). These data taken together indicate that *RAS* proteins interact directly with adenylate cyclase.

2.5 Identification of a Protein Domain of *RAS* Essential for Activation of Adenylate Cyclase

In an attempt to identify the protein domain of *RAS* proteins which bind to and activate the "target" of *RAS*, Sigal and coworkers analyzed the biological properties of several *RAS* mutant proteins (SIGAL et al. 1986a). They reasoned that mutants which abolish the ability of *RAS* to activate its target might not affect the *RAS* protein's ability to bind GTP, hydrolyze GTP, and localize to the plasma membrane. Several point mutations were found which did not effect the intrinsic biochemical properties of the *RAS* proteins but destroyed the ability of *RAS* to activate adenylate cyclase in yeast. All of these mutations lead to amino acid substitutions at positions 32–40 of the H-*ras* protein (SIGAL et al. 1986a, SIGAL et al. 1986b). Also these mutants could not function in mammalian cells, suggesting they had lost their ability to interact and activate their mammalian target. SIGAL et al. were careful to point out that other explanations for these data exist; for example mutations in the 32–40 region might alter the conformation of a distant region which constitutes the actual domain which binds to the target. Nonetheless, it is interesting that the amino acids 32–40 domain of H-*ras* is highly conserved in evolution. The sequence of this domain is invariant in all known *RAS* proteins from yeast to man (SIGAL et al. 1986a). The high degree of amino acid sequence conservation of this domain might reflect the conservations of *RAS* protein-protein interaction during evolution. If this is true, one might expect the protein domain of the yeast adenylate cyclase to which *ras* binds to have been structurally conserved in evolution.

2.6 The *ras* Proteins Do Not Activate Adenylate Cyclase in Vertebrates

The finding that the human and yeast *RAS* proteins can activate the yeast adenylate cyclase prompted the investigation which asked if the vertebrate adenylate cyclase was regulated by *ras* proteins. BIRCHMEIER and coworkers observed that H-$ras^{val\,12}$ proteins injected into *Xenopus* oocytes induced maturation (BIRCHMEIER et al. 1985). Oocyte maturation can be stimulated by agents which decrease cAMP levels as well as by an insulin-activated pathway which does not lead to changes in cAMP levels. The induction of maturation caused by injection of *ras* proteins did not accompany any detectable changes in cAMP levels, and this work concluded that *ras* does not regulate vertebrate adenylate cyclase. Subsequently, it was observed that insulin-induced oocyte maturation could be blocked by antibodies against *ras* proteins, suggesting that *ras* functions in the insulin-dependent oocyte maturation pathway (KORN et al. 1987).

Thus, although there is considerable evidence to suggest there is functional homology between the human and yeast *RAS* proteins, the immediate biochemical function of the yeast *RAS* proteins (activation of adenylate cyclase) does not appear to be conserved in evolution. How might this apparent paradox be explained? First *RAS* proteins might have two functions (and hence two targets) in yeast, and one of the functions (activation of adenylate cyclase) might have been lost in the course of evolution. Indeed, there is evidence for an alternate function for *RAS* in yeast, but the direct target has not been identified (TODA et al. 1987b). Secondly, the domain of adenylate cyclase to which *RAS* proteins bind might be conserved in evolution but be found in mammals in a protein with a catalytic function other than adenylate cyclase.

2.7 The Bovine GTPase-Activating Protein Is Homologous in Structure to a Domain of the Yeast Adenylate Cyclase

Recently, a GTPase-activating protein (GAP) has been found in *Xenopus* oocytes and mammalian cell extracts that stimulate the GTPase activity of normal but not oncogenic H-*ras* proteins (Trahey and McCormick 1987; Vogel et al. 1988). As discussed above in Sect. 2.5, H-*ras* mutant proteins with amino acid substitutions in the 32–40 region cannot stimulate the yeast adenylate cyclase. These H-*ras* mutants also cannot interact with the bovine GAP, suggesting that a single domain of the H-*ras* protein is required for interaction of both GAP and the yeast adenylate cyclase (Vogel et al. 1988). The bovine GAP can also stimulate the GTPase activity of the yeast *RAS2* protein (J. Gibbs, personal communication). Thus, the GAP, like the yeast adenylate cyclase, can interact directly with *RAS* proteins from yeast and mammals. Also, it is noteworthy that the purified yeast adenylate cyclase does not appear to contain GAP activity (J. Field and R.-M. Ballister, personal communication). Interestingly, GAP and the yeast adenylate cyclase share amino acid sequence homology (Vogel et al. 1988). The entire bovine GAP sequence shows 16% identity when compared to amino acid positions 425–1462 of the yeast adenylate cyclase noncatalytic domain. Taking into account conservative amino acid substitution, this homology increases to 40% (Vogel et al. 1988). The precise role GAP plays in the *ras* growth control pathway is not yet clear. Nonetheless, the comparison of the GAP and adenylate cyclase sequences might point to regions in the GAP where structure has been conserved in evolution; these regions many have been conserved because they have functionally essential roles.

3 *CDC25* Function in Yeast and Its Interaction with *RAS* Proteins

3.1 *CDC25* Regulates Growth in Yeast

cdc25^{ts} strains have many of the defects characteristic of strains with defects in the *RAS*/cAMP pathway. *cdc25*^{ts} strains arrest in G1 in a state resembling that of nutritionally starved cells (Pringle and Hartwell 1981). Like nutritionally starved cells, *cdc25*^{ts} cells at the restrictive temperature present a marked decrease in stable RNA synthesis (Martegani et al. 1984). Also, *cdc25*^{ts} arrested cells have an increased synthesis of heat-shock proteins, a feature generally observed in growth-arrested cells (Iida and Yahara 1984). *cdc25*^{ts} arrested cells, like growth-arrested cells, are particularly resistant to heat shock and accumulate large stores of glycogen (Martegani et al. 1986a). *cdc25*^{ts}/*cdc25*^{ts} homozygous diploid strains, like *cyr1*^{ts}/*cyr1*^{ts} homozygous diploid strains, sporulate without being starved for nitrogen (Shilo et al. 1978). All these observations taken together prompted several laboratories to investigate the role *CDC25* plays in cAMP mediated growth control.

3.2 *CDC25* Regulates cAMP Levels in Yeast

In a clear demonstration of the involvement of *CDC25* in the cAMP pathway, MARTEGANI et al. (1986a) and CAMONIS et al. (1986) showed that in cAMP-permeable strains, exogenously added cAMP can suppress the *cdc25*^{ts} growth-arrest defect. While exogenous cAMP can suppress the defect of both the *cdc25-1* and *cdc25-5* alleles, examination of intracellular cAMP levels suggested *CDC25* may have a complex function. At the nonpermissive temperature, cAMP levels drop dramatically in *cdc25-5* strains (CAMONIS et al. 1986), whereas no change in cAMP levels is observed in *cdc25-1* strains (MARTEGANI et al. 1986a). Despite the high levels of cAMP in *cdc25-1* strains, they arrest in G1 at the nonpermissive temperature, suggesting that *CDC25* has a growth-promoting role other than the control of cAMP levels. It is clear however, that this putative second function of *CDC25* is not essential if cAMP levels are artificially elevated. First, the ts defect of both *cdc25-1* and *cdc25-5* strains can be suppressed by exogenous cAMP (MARTEGANI et al. 1986a). Secondly, the lethality resulting from disruption of the *CDC25* locus can be suppressed by overexpression of the yeast adenylate cyclase or introduction of *RAS2*^{val}. These observations suggest that *CDC25* functions upstream of *RAS* and that if *CDC25* has a function other than regulation of *RAS*, it is not essential in the presence of high levels of intracellular cAMP.

Recent evidence has suggested that the *CDC25* gene product is involved in the signal transmission pathway of the glucose-induced cAMP signal in yeast. Cells arrested by glucose starvation when refed with glucose exhibit a dramatic but transient increase in intracellular cAMP levels (MBONYI et al. 1988). This glucose-induced cAMP peak is deficient in yeast strains containing a ts *cdc25* allele at the nonpermissive temperature.

3.3 *CDC25* Regulates *RAS* Function in Yeast

More recent experiments indicate that the activity of the *RAS* proteins is controlled by the product of the *CDC25* gene (ROBINSON et al. 1987; BROEK et al. 1987; DANIEL et al. 1987). The GTP responsiveness of adenylate cyclase is abnormal in cell membranes from strains lacking *CDC25* (BROEK et al. 1987; ROBINSON et al. 1987). Membranes prepared from strains lacking *CDC25* exhibit less than 10% of the basal Mg^{2+}-dependent adenylate cyclase activity observed in strains containing *CDC25* (Table 1, lines 9, 10). However, the addition of Gpp(NH)p to membranes from strains lacking *CDC25* stimulated the Mg^{2+}-dependent adenylate cyclase activity to a level similar to that observed in strains containing *CDC25* (BROEK et al. 1987). Thus, the *RAS* proteins in *CDC25*-deficient strains are capable of function in vitro when a nonhydrolyzable GTP analog is present. However, the defect in basal adenylate cyclase activity suggests that the *RAS* proteins in the isolated membranes are inactive (GDP-bound) in a *cdc25*$^-$ strain. These observations suggest the *CDC25* protein may activate *RAS* proteins by promoting GDP/GTP exchange of the *RAS* proteins (BROEK et al. 1987; ROBINSON et al. 1987). In the absence of *CDC25*, the *RAS* proteins would be bound primarily to GDP, and would therefore be inactive.

The *RAS2*^{val19} mutation results in the production of a *RAS2* protein with a markedly reduced GTPase activity. Thus, the *RAS2*^{val19} protein found in growing yeast cells would be expected to be bound to GTP more often than their wild-type counterparts. Indeed, less than 1% of wild-type *RAS2* protein is found bound to GTP in vivo whereas 33% of *RAS2*^{val19} protein was found bound to GTP (GIBBS et al. 1987). *RAS* proteins bound to GTP (but not those bound to GDP) activate adenylate cyclase (FIELD et al. 1987). Thus, the *RAS2*^{val19} mutation leads to overproduction of cAMP. A consequence of this is that *RAS2*^{val19} cells are unable to sense the absence of nutrients and therefore do not arrest in G1 when starved (TODA et al. 1985).

The growth-arrest defect of *RAS2*^{val19} strains can be partially attenuated by disruption of the *CDC25* gene. Wild-type cells which are growth arrested are resistant to heat shocks. *RAS2*^{val19} strains relative to wild-type strains are very sensitive to heat shocks. The heat-shock sensitivity of *RAS2*^{val19} strains is lessened by disruption of *CDC25* (BROEK et al. 1987). This observation demonstrated that *RAS*^{val19} is slightly dependent on *CDC25* function. This dependence can be observed by analysis of the adenylate cyclase activity in membranes from strains lacking *CDC25* and containing *RAS2*^{val19}. The basal Mg^{2+}-dependent adenylate cyclase activity of *RAS2*^{val19} membranes is elevated relative to wild-type membranes and cannot be stimulated by the addition of Gpp(NH)p. In contrast *cdc25*$^{-}$, *RAS*^{val19} membranes exhibit a decrease in basal Mg^{2+}-dependent adenylate cyclase activity relative to a *RAS2*^{val19} strain and this activity can be stimulated by Gpp(NH)p (Table 1, lines 10, 11; BROEK et al. 1987).

The proposed model of *CDC25* function suggests that the *CDC25* gene product is not only necessary for the transmission of the glucose-induced signal but that *CDC25* can affect the amplitude of this signal. In support of this suggestion, mutations in *CDC25* have been identified which cause cells to be unable to arrest properly in G1 when starved for glucose (BROEK et al. 1987; MARTEGANI et al. 1986b). Yeast harboring the mutant *CDC25*^{HS} gene, even after starvation, have an increased sensitivity to heat shocks. This inability to arrest properly is presumably due to increased intracellular cAMP levels. Thus it appears the *CDC25* protein can affect the amplitude of the cAMP signal.

3.4 *CDC25* May Interact Directly with Yeast and Human *RAS* Proteins

POWERS and coworkers, studying a *RAS2* point mutant (*RAS2*^{ala22}), have obtained evidence suggesting a direct interaction between the *CDC25* gene product and *RAS2* proteins. Wild-type yeast transformed with a plasmid containing *RAS2*^{ala22} exhibit growth arrest in G1 of the cell cycle at 37 °C but grew well at 30 °C. This growth-arrest defect can be suppressed by overexpression of *CDC25*, but only in the presence of a wild type *RAS* gene (S. POWERS, personal communication). POWERS and coworkers propose a model to explain these data. This model postulates that *CDC25* protein interacts directly but transiently with *RAS* proteins bound to GDP, and causes GDP dissociation. The interaction between *RAS* and *CDC25* is completed when *RAS* proteins bind GTP. This mechanism of catalysis has been proposed by Stryer for the activation of transducin by photorhodopsin (STRYER 1986). According to the model,

RAS2^{ala22} has a reduced affinity for guanine nucleotides and binds stably to *CDC25* proteins, thereby preventing *CDC25* from acting on the wild-type *RAS* protein. The expression of H-*ras* mutants with reduced affinity for guanine nucleotides are able to block growth in *S. cerevisiae* (SIGAL et al. 1986b). Sigal and coworkers proposed that inhibition of growth was caused by the H-*ras* mutants forming a dead-end complex with a protein with which wild-type *ras* proteins interact in signal transduction. POWERS and coworkers showed that the human H-*ras*^{ala15} (analogous to the *RAS2*^{ala22}) causes a growth-arrest defect in yeast which can be suppressed by overexpressing *CDC25* (S. POWERS, personal communication). This observation suggests the H-*ras* protein is able to deplete the endogenous supply of *CDC25*, thus the protein-protein interaction involving *RAS* and *CDC25* may be conserved in evolution.

4 References

Barbicid M (1987) *RAS* gene. Annu Rev Biochem 56: 779–828

Birchmeier C, Broek D, Wigler M (1985) *RAS* proteins can induce meiosis in Xenopus oocytes. Cell 43: 615–621

Boutelet F, Petitjean A, Hilger F (1985) Yeast *CDC35* mutants are defective in adenylate cyclase and are alleleic with *CYR1* mutants while *CAS1*, a new gene, is involved in the regulation of adenylate cyclase. EMBO J 4: 2635–2642

Broek D, Samiy N, Fasano O, Fujiyama A, Tamanoi F, Northup J, Wigler M (1985) Differential activation of yeast adenylate cyclase by wild-type and mutant *ras* proteins. Cell 41: 763–769

Broek D, Toda T, Michaeli T, Levin L, Birchmeier C, Zoller M, Powers S, and Wigler M (1987) The *S. cerevisiae CDC25* gene product regulates the *RAS*/adenylate cyclase pathway. Cell 48: 789–799

Camonis JH, Kalikine M, Bernard G, Garreau H, Boy-Marcotte E, Jacquet M (1986) Characterization, cloning and sequence analysis of the *CDC25* gene which controls the cyclic AMP level of *Saccharomyces cerevisiae*. EMBO J 5: 375–380

Cannon J, Tatchell K (1987) Characterization of the *Saccharomyces cerevisiae* gene encoding subunits of cyclic AMP-dependent protein kinase. Mol Cell Biol 7: 2653–2663

Cannon JF, Gibbs JB, Tatchell K (1986) Suppressors of the *ras2* mutations of *S. cerevisiae*. Genetics 113: 247–264

Casperson G, Walker N, Brasier A, Bourne H (1983) A guanine nucleotide-sensitive adenylate cyclase in the yeast *Saccharomyces cerevisiae*. J Biol Chem 258: 7911–7914

Casperson G, Walker N, Bourne H (1985) Isolation of the gene encoding adenylate cyclase in the yeast *Saccharomyces cereviciae*. Proc Natl Acad Sci USA 82: 5060–5063

Daniel J, Becker J, Enari E, Levitzki A (1987) The activation of adenylate cyclase by guanyl nucleotides in *Saccharomyces cerevisiae* is controlled by the *CDC25* start gene product. Mol Cell Biol 7: 3857–3861

DeFeo-Jones D, Scolnick EM, Zoller R, Dhar R (1983) *ras*-related gene sequences identified and isolated from *Saccharomyces cerevisiae*. Nature 306: 707–709

DeFeo-Jones D, Tatchell K, Robinson LC, Sigal I, Vass W, Lowy DR, Scolnick EM (1985) Mammalian and yeast *ras* gene products: biological function in their heterologous systems. Science 228: 179–184.

Deschenes RJ, Broach JR (1987) Fatty acylation is important but not essential for *Saccharomyces cerevisiae RAS* function. Mol Cell Biol 7: 2344–2351

Esposito RE, Klapholz S (1981) Meiosis and ascospore development. In: Strathern JN, Jones EW, Broach JR (eds) The molecular biology of the yeast *Saccharomyces*: Life cycle and inheritance. Cold Spring Harbor Laboratory, Cold Spring Harbor, pp 211–287

Field J, Broek D, Kataoka T, Wigler M (1987) Guanine nucleotide activation of, and competition between, *RAS* proteins from *Saccharomyces cerevisiae*. Mol Cell Biol 7: 2128–2133

Gibbs JB, Sigal IS, Poe M, Scolnick EM (1984) Intrinsic GTPase activity distinguishes normal and oncogenic *ras* p21 molecules. Proc Natl Acad Sci USA 81: 5704–5708

Gibbs JB, Schaber MD, Marshall MS, Scolnick EM, Sigal IS (1987) Identification of guanine nucleotides bound to *ras*-encoded proteins in growing yeast cells. J Biol Chem 262: 10426–10429

Gilman A (1987) G proteins: transducers of receptor-generated signals. Annu Rev Bioch 56: 615–649

Hartwell LH (1974) *Saccharomyces cerevisiae* cell cycle. Bacteriol Rev 38: 164–198

Hixson CS, Krebs EG (1980) Characterization of a cyclic AMP-binding protein from baker's yeast. Identification as a regulatory subunit of cyclic AMP dependent protein kinase. J Biol Chem 225: 2137–2145

Iida H, Yahara I (1984) A heat shock resistant mutant of *Saccharomyces cerevisiae* shows constitutive synthesis of two heat shock proteins and altered growth. J Cell Biol 99: 1441–1450

Johnson K, Cameron S, Toda T, Wigler M, Zoller M (1987) Expression in *E. coli* of BCY 1, the regulatory subunit of the cAMP dependent protein kinase from *Saccharomyces cerevisiae*. J Biol Chem 255: 2137–2145

Johnston GC, Pringle JR, Hartwell LH (1977) Coordination of growth with cell division in the yeast *Saccharomyces cerevisiae*. Exp Cell Res 105: 79–98

Kataoka T, Powers S, McGill C, Fasano O, Strathern J, Broach J, Wigler M (1984) Genetic analysis of yeast *RAS1* and *RAS2* genes. Cell 37: 437–445

Kataoka T, Powers S, Cameron S, Fasano O, Goldfarb M, Broach H, Wigler M (1985a) Functional homology of mammalian and yeast *RAS* genes. Cell 40: 19–26

Kataoka T, Broek D, Wigler M (1985b) DNA sequence and characterization of the *S. cerevisiae* gene encoding adenylate cyclase. Cell 43: 493–505

Korn L, Siebel C, McCormick F, Roth R (1987) *Ras* p21 as a potential mediator of insulin action in Xenopus oocytes. Science 236: 840–843

Kuret J, Johnson K, Nicolette C, Zoller M (1988) Metagenesis of the regulatory subunit of yeast cAMP-dependent protein kinase. J Biol Chem 263: 9149–9151

Lillie SH, Pringle JR (1980) Reserve carbohydrate metabolism in *Saccharomyces cerevisiae*: responses to nutrient limitation. J Bacteriol 143: 1384–1394

Martegani E, Vanoni M, Baroni M (1984) Macromolecular synthesis in the cell cycle mutant *cdc25* of budding yeast. Eur J Biochem 144: 205–210

Martegani E, Baroni M, Wanoni M (1986a) Interaction of cAMP with the *CDC25*-mediated step in the cell cycle of *Saccharomyces cerevisiae*. Exp Cell Res 544–548

Martegani E, Baroni M, Frascotti G, Alberghina L (1986b) Molecular cloning and transcriptional analysis of the start gene *CDC25* of *Saccharomyces cerevisiae*. EMBO J 5: 2363–2369

Matsumoto K, Uno I, Oshima Y, Ishikawa T (1982) Isolation and characterization of yeast mutants deficient in adenylate cyclase and cAMP-dependent protein kinase. Proc Natl Acad Sci USA 79: 2355–2359

Matsumoto K, Uno I, Ishikawa T (1983) Control of cell division in *Saccharomyces cerevisiae* mutants defective in adenylate cyclase and cAMP-dependent protein kinase. Exp Cell Res 146: 151–161

Matsumoto K, Uno I, Ishikawa T (1984) Identification of the structural gene and nonsense allels for adenylate cyclase in *Saccharomyces cerevisiae*. J Bacteriol 157: 277–282

Matsumoto K, Uno I, Ishikawa T (1985) Genetic analysis of the role of cAMP in yeast. Yeast 1: 15–24

Mbonyi K, Beullens M, Detremeriek K, Geerts L, Thevelein J (1988) Requirement of one functional *RAS* gene and inability of an oncogenic *ras* variant to mediate the glucose-induced cyclic AMP signal in the yeast Saccharomyces cerevisiae. Mol Cell Biol 8: 3051–3057

McGrath JP, Capon DJ, Goeddel DV, Levinson AD (1984) Comparative biochemical properties of normal and activated human *ras* p21 protein. Nature 310: 644–649

Papageorge A, Lowy DR, Scolnick EM (1982) Comparative biochemical properties of p21 *ras* molecules coded for viral and cellular *ras* genes. J Virol 44: 509–519

Powers S, Kataoka T, Fasano O, Goldfarb M, Strathern J, Broach J, Wigler M (1984) Genes in *S. cerevisiae* encoding proteins with domains homologous to the mammalian *ras* proteins. Cell 36: 607–612

Powers S, Michaelis S, Broek D, Santa Anna-A S, Field J, Herskowitz I, Wigler M (1986) *RAM*, a gene of yeast required for a functional modification of *RAS* proteins and for production of mating pheromone a-factor. Cell 47: 413–422

Pringle JR, Hartwell LH (1981) The *Saccharomyces cerevisiae* cell cycle. In: Strathern JN, Jones EW, Broach JR (eds) The molecular biology of the yeast *Saccharomyces*: Life cycle and inheritance. Cold Spring Harbor Laboratory, Cold Spring Harbor, pp 97–142

Robinson LC, Gibbs JB, Marshall MS, Sigal IS, Tatchell K (1987) *CDC25*: A component of the *RAS*-adenylate cyclase pathways in *Saccharomyces cerevisiae*. Science 235: 1218–1221

Shilo V, Simchen G, Shilo B (1978) Initiation of meiosis in cell cycle initiation mutants of *Saccharomyces cerevisiae*. Exp Cell Res 112: 241–249

Shimizu K, Goldfarb M, Suard Y, Perucho M, Li Y, Kamata T, Feramisco J, Stavnezer E, Fogh J, Wigler M (1983) Three human transforming genes are related to the viral *ras* oncogenes. Proc Natl Acad Sci USA 80: 2112–2116

Shin D-Y, Uno I, Ishikawa T (1987) Control of the G1-G_0 transition and G_0 protein synthesis by cyclic AMP in *Saccharomyces cerevisiae*. Curr Genet 12: 577–582

Sigal IS, Gibbs JB, D'Alonzo JS, Temeles GL, Wolanski BS, Socher SH, Scolnick EM (1986a) Mutant *ras*-encoded proteins with altered nucleotide binding exert dominant biological effects. Proc Natl Acad Sci USA 83: 952–956

Sigal IS, Gibbs JB, D'Alonzo JS, Scolnick EM (1986b) Identification of effector residues and a neutralizing epitope of Ha-*ras*-encoded p21. Proc Natl Acad Sci USA 83: 4725–4729

Stryer L (1986) Cyclic GMP cascade of vision. Annu Rev. Neurosci 9: 87–119

Sweet R, Yokoyama S, Kamata T, Feramisco J, Rosenberg M, Gross M (1984) The product of *ras* is a GTPase and the T24 oncogenic mutant is deficient in this activity. Nature 311: 273–275

Tamanoi F, Walsh M, Kataoka T, Wigler M (1984) A product of yeast *RAS2* gene is a guanine nucleotide binding protein. Proc Natl Acad Sci USA 81: 6924–6929

Taparowsky E, Suard Y, Fasano O, Shimizu K, Goldfarb M, Wigler M (1982) Activation of the T24 bladder carcinoma transforming gene is linked to a single amino acid change. Nature 300: 762–765

Taparowsky E, Shimizu K, Goldfarb M, Wigler M (1983) Structure and activation of the human N-*ras* gene. Cell 34: 581–586

Tatchell K, Chaleff D, Defeo-Jones D, Scolnick E (1984) Requirement of either of a pair of *ras* related genes of *Saccharomyces cerevisiae* for spore viability. Nature 309: 523–527

Tatchell K, Robinson LC, Breitenbach M (1985) *RAS2* of *Saccharomyces cerevisiae* is required for gluconeogenic growth and proper response to nutrient limitation. Proc Natl Acad Sci USA 82: 3785–3789

Temeles GL, Gibbs JB, D'Alonzo JS, Sigal IS, Scolnick EM (1984) Yeast and mammalian *ras* proteins have conserved biochemical properties. Nature 313: 700–703

Toda T, Uno I, Ishikawa T, Powers S, Kataoka T, Broek D, Cameron S, Broach J, Matsumoto K, Wigler M (1985) in yeast, *RAS* proteins are controlling elements of adenylate cyclase. Cell 40: 27–36

Toda T, Cameron S, Sass P, Scott J, McMullen B, Hurwitz M, Krebs E, Wigler M (1987a) Cloning and characterization of *BCY1*, a locus encoding a regulatory subunit of the cyclic AMP-dependent protein kinase in *Saccharomyces cerevisiae*. Mol Cell Biol 7: 1371–1377

Toda T, Broek D, Field J, Michaeli T, Cameron S, Nikawa J, Sass P, Birchmeier C, Powers S, Wigler M (1987b) Exploring the function of the *RAS* oncogenes by studying the yeast editor S. Aaronson *Saccharomyces cerevisiae*. In: Oncogenes and cancer. Japan Sci Soc Press, Tokyo/VNU, Utrecht, pp 253–260

Toda T, Sass P, Zoller M, Wigler M (1987c) Three different genes in S. cerevisiae erode the catalytic subunits of the subunits of the cAMP dependent protein kinase. Cell 50: 277–287

Trahey M, McCormick F (1987) A cytoplasmic protein regulates the function of normal *ras* p21 by stimulating GTPase activity, but does not affect oncogenic mutants. Science 238: 542–545

Uno I, Matsumoto K, Ishikawa T (1982) Characterization of cyclic AMP-requiring yeast mutants altered in the regulatory subunit of protein kinase. J Biol Chem 257: 14110–14115

Uno I, Matsumoto K, Aduchi K and Ishikawa T (1984a) Genetic and biochemical evidence that trehalase is a substrate of cAMP-dependent protein kinase in yeast. J Biol Chem 258: 10867–10872

Uno I, Mitsuzawa H, Matsumoto K, Tanaka K, Ushima I, Ishikawa T (1985) Reconstitution of the GTP-dependent adenylate cyclase from products of the yeast *CYRI* and RAS2 genes in *Escherichia coli.* Proc Natl Acad Sci USA 82: 7855–7859

Uno I, Matsumoto K, Adachi K, Ishikawa T (1984b) Characterization of cAMP-requiring mutants altered in the catalytic subunit of protein kinase. J Biol Chem 259: 12508–12513

Vogel U, Dixon R, Schaber M, Diehl R, Marshall M, Scolnick E, Sigal I, Gibbs J (1988) Cloning of bovine GAP and its interaction with oncogenic *ras* p21. Nature 335: 90–93

Subject Index

MIX
Papier aus verantwortungsvollen Quellen
Paper from responsible sources
FSC® C105338

If you have any concerns about our products,
you can contact us on
ProductSafety@springernature.com

In case Publisher is established outside the EU,
the EU authorized representative is:
Springer Nature Customer Service Center GmbH
Europaplatz 3, 69115 Heidelberg, Germany

Printed by Libri Plureos GmbH
in Hamburg, Germany